TROUBLES WITH ...

New Directions in Anthropology
General Editor: Jacqueline Waldren, *Institute of Social Anthropology, University of Oxford*

Volume 1   *Coping with Tourists: European Reactions to Mass Tourism*
            Edited by Jeremy Boissevain

Volume 2   *A Sentimental Economy: Commodity and Community in Rural Ireland*
            Carles Salazar

Volume 3   *Insiders and Outsiders: Paradise and Reality in Mallorca*
            Jacqueline Waldren

Volume 4   *The Hegemonic Male: Masculinity in a Portuguese Town*
            Miguel Vale de Almeida

Volume 5   *Communities of Faith: Sectarianism, Identity, and Social Change on a
            Danish Island*
            Andrew S. Buckser

Volume 6   *After Socialism: Land Reform and Rural Social Change in Eastern Europe*
            Edited by Ray Abrahams

Volume 7   *Immigrants and Bureaucrats: Ethiopians in an Israeli Absorption Center*
            Esther Hertzog

Volume 8   *A Venetian Island: Environment, History and Change in Burano*
            Lidia Sciama

Volume 9   *Recalling the Belgian Congo: Conversations and Introspection*
            Marie-Bénédicte Dembour

Volume 10   *Mastering Soldiers: Conflict, Emotions, and the Enemy in an Israeli
            Military Unit*
            Eyal Ben-Ari

Volume 11   *The Great Immigration: Russian Jews in Israel*
            Dina Siegel

Volume 12   *Morals of Legitimacy: Between Agency and System*
            Italo Pardo

Volume 13   *Academic Anthropology and the Museum*
            Edited by Mary Bouquet

Volume 14   *Simulated Dreams: Israeli Youth and Virtual Zionism*
            Haim Hazan

Volume 15   *Defiance and Compliance: Negotiating Gender in Low-Income Cairo*
            Heba Aziz Morsi El-Kholy

Volume 16   *Troubles with Turtles: Cultural Understandings of the Environment on a
            Greek Island*
            Dimitrios Theodossopoulos

Volume 17   *Rebordering the Mediterranean: Boundaries and Citizenship in Southern
            Europe*
            Liliana Suarez-Navaz

Troubles with Turtles

Cultural Understandings of the Environment on a Greek Island

Dimitrios Theodossopoulos

Berghahn Books
New York • Oxford

First published in 2003 by **Berghahn Books**
www.berghahnbooks.com

First paperback edition published in 2005

Library of Congress Cataloging-in-Publication Data

Theodossopoulos, Dimitrios.
Troubles with turtles : cultural understandings of the environment on a Greek
    island / Dimitrios Theodossopoulos.
        p. cm. -- (New directions in anthropology ; v. 16)
Includes bibliographical references.
    ISBN 1-57181-596-1 (alk. paper) -- ISBN 1-57181-697-6 (pb.: alk. paper)
    1. Nature conservation--Social aspects--Greece--Zakynthos. 2. Nature
conservation--Economic aspects--Greece--Zakynthos. 3. Human ecology--
Greece--Zakynthos. I. Title. II. Series.

QH77.G8 T54 2002
333.95'16'09495--dc21

                                                                2002025598

British Library Cataloguing in Publication Data
A catalogue record for this book is available from
the British Library.

Printed in the United States on acid-free paper

Figure 1 Newly hatched sea turtles emerging on the beach

CONTENTS

List of Figures   viii

Acknowledgements   x

A Note on Transliteration   xii

   1. Introduction   1

   2. Vassilikos: past, present and turtle troubles   13

   3. Conservation and the value of land   29

   4. 'Both tourism and farming jobs involve struggle' 49

   5. Gendered labour in the olive harvest   67

   6. Ordering animals about   89

   7. Classifying the wild   111

   8. Unlawful hunting   139

   9. Relating to the 'natural' world   161

References   179

Index   193

LIST OF FIGURES

1 Newly hatched sea turtles emerging on the beach   v
2 Vassilikos road sign   16
3 Map of Zakynthos   21
4 Michalis's old donkey before engaging in a new career in tourism   96
5 Mimis, one of the older flock owners, with his flock and sheep fold   101
6 Lefteris in the struggle of animal care   106

To the people of Vassilikos,
those who resist conservation,
and those who don't.

ACKNOWLEDGEMENTS

Twelve years ago I arrived in Zakynthos intending to study environmental politics, attracted by the conflict between groups of environmentalists and the indigenous communities. At the time, I was planning to focus on the conflict itself and the growing movement of environmentalism in Greece. I can now only speculate about the exact reasons that led me to shift my original emphasis from environmental politics in general towards the indigenous community of Vassilikos in particular. It may have been my deep conviction that only a carefully conducted study of the culture of those people who resist conservation would lead to an understanding of their actual resistance towards conservation. It may also have been an idealistic desire to speak for those who had the greatest difficulty representing themselves, the more inarticulate and the less privileged. In both cases, I was influenced by the unique perspective of my discipline (anthropology), its respect for local knowledge and its moral commitment to representing misrepresented indigenous voices.

Studying Vassilikos was a hard task. It took time to earn the trust of the inhabitants and data on what soon became my new main topic – the indigenous human-environmental relationship – was not easy to collect. During my days in Zakynthos, I made my life harder by denying myself most of the pleasures that the island had to offer – the beautiful beaches, the tourist amenities. I now regret that I followed this self-punitive approach, because I know that the people of Vassilikos would have helped me anyway, even if I had made my life more comfortable. I devote this book to them, not simply out of professional courtesy, but because they provided me with the initial impetus to write *this* kind of book.

In the ethnography that follows, I have avoided referring to actual names. Where that was not possible, I have consistently applied the pseudonym 'Dionysis', the most commonplace and representative male Zakynthian and Vassilikiot name. In fact, whenever the name Dionysis appears in my text, it is *always* a pseudonym. In some other instances, however, I name my respondents with their actual first names, such as my references to Lefteris, my adoptive father in the field. These are the cases where I know that the people in question would like their real names to be included. Several Vassilikiot men and women expressed their explicit or implicit desire 'to be in' my book. I believe that I have fulfilled their wishes by letting them speak in their own words as far as possible.

Speaking of actual names, I should like to express my enormous dept to Lefteris Gianoulis, Dionysis Tsilimigras and their families. They were patient teachers in every aspect of Vassilikos culture and they remained close to me while I was in Vassilikos, during good times and bad times alike. I should also like to thank my teachers in academia, Peter Loizos and Paul Richards; the first taught me the art of writing anthropology (and how to teach this art to my students); the second initiated me in the anthropology of the environment. During the writing of this book, I received encouragement and moral support from several fellow-anthropologists. For their valuable comments and suggestions I wish to thank Elisabeth Kirtsoglou, Veronica Strang, Margaret Kenna, Charles Stewart, Allen Abramson, Mark Jamieson, Pola Boussiou, Marios Sarris, Panos Panopoulos, Akis Papataxiarchis, Keith Brown and the editor of this series, Jacqueline Waldren.

I wish to offer many thanks to my proof-readers, Valerie Nunn and Roger Smedley, for the time and effort they spent on my manuscript. I am also grateful to Veronica Strang and Chris Stray for their creative title-suggestions and George Chiras for his photographs. I cannot forget the help I received from two people who collected rare library material for me, my mother Afrodite Theodossopoulou, and her sister Tasia Kolokotsa: surprisingly, they derive pleasure from calling themselves my 'research assistants' and they certainly deserve this title! Once again, I thank my partner Elisabeth Kirtsoglou for her everyday patience and support in academic matters and beyond. Last, but greater than all, is my gratitude to my father Andreas Theodossopoulos, whose passionate encouragement made this long journey feasible. Without him, I would not even have started.

I should acknowledge, finally, that some of the topics I address in this book have already been discussed in one form or another in other publications. In particular, I have examined environmental politics in Zakynthos in two articles, one in the *Journal of Mediterranean Studies* (7,2: 250-67) and one in the volume edited by Dawn Chatty and Marcus Colchester (*Conservation and Indigenous Mobile Peoples*). The human-animal relationship in Vassilikos was explored in a volume edited by John Knight (*Animals in Person*), while the legal parameters of the indigenous relationship with the land were examined in a volume which I edited with Allen Abramson (*Land, Law and Environment*). The gendered division of labour in the olive harvest was discussed in an article published in the *Journal of the Royal Anthropological Institute* (5,4: 611-626). My earlier accounts of these topics have now been revised and completely rewritten, and although some of the main arguments inevitably remain the same, they are now used in a more extended and organised attempt to draw a complete and coherent picture of the human environmental relationship in Vassilikos.

A Note on Transliteration

Indigenous terms that are used more directly in the text are transliterated according to the system presented below with the main exception of names and placenames. Considering that 'there is no single system of transliteration that is accepted by all scholars of modern Greek' (Dubisch 1995a: xv), I have attempted to achieve 'a compromise between the phonology and orthography of modern Greek' (Stewart 1991: xi). In particular, I have chosen to transliterate the digraphs ει, αι, οι and ου with ei, ai oi and ou respectively, to account more closely for Greek orthography. In most other respects I have complied with the style recommended by the *Journal of Modern Greek Studies*.

α	a	[alpha]	αι	ai	
β	v	[beta]	ει	ei	
γ	g	[gamma]	οι	oi	
δ	dh	[delta]	ου	ou	
ε	e	[epsilon]	αυ	af (before voiceless consonants)	
ζ	z	[zeta]		av (before vowels/voiced consonants)	
η	i	[eta]	ευ	ef (before voiceless consonants)	
θ	th	[theta]		ev (before vowels, voiced consonants)	
ι	i	[iota]	γχ, γγ	g	
κ	k	[kappa]	ντ	d (initial), nt (medial)	
λ	l	[lambda]	μπ	b (initial), mp (medial)	
μ	m	[mu]			
ν	n	[nu]			
ξ	x	[xi]			
ο	o	[omicron]			
π	p	[pi]			
ρ	r	[rho]			
σ	s	[sigma]			
τ	t	[tau]			
υ	y	[upsilon]			
φ	f	[phi]			
χ	kh	[chi]			
ψ	ps	[psi]			
ω	o	[omega]			

1
INTRODUCTION

Have a look around you… Do you see these beautiful fields, the beautiful nature? This is our land. We've been working this land since we were children. Once upon a time, we used to have the landlords telling us what to do. Then, with tourism, just as we lifted our heads up, the ecologists started talking about the turtle. They told us we cannot build on our land, we cannot make any progress on it. They say they care about the earth. What do they know? They say we should protect the turtle. Who is going to protect us from the turtle? What good does the turtle do? It only brings troubles.

What good can a turtle do a human? Why do we have to pay so much attention to them? Here, the turtle has caused great harm to the people. It went against the interests of the people. I'll tell you something. If you came here to advance the cause of the turtle, you'd better go away. But if you came to write about the people, then I'll tell you all I know…

This monograph is an in-depth study of the relationship some people have with the natural world. The inhabitants of a Greek island community, who are simultaneously farmers and tourist entrepreneurs, are involved in a bitter environmental dispute concerning the imposition of conservation regulations on the local environment. Their community, *Vassilikos*, is located in the southwest corner of the island of Zakynthos, a major tourist destination in the Mediterranean. Vassilikos's prolonged and persistent resistance to environmental conservation has inspired this monograph.

Twelve years ago I began my investigation of the human environmental relationship in Vassilikos, considering that an anthropological study of the indigenous culture was a necessary prerequisite for understanding local environmental disputes. This undertaking was intended not merely as an addition to the anthropological literature – a detailed ethnography on the human environmental relationship in southeast Europe – but also as an example of how anthropology can contribute to a socially informed understanding of environmental disputes. It has been noted that at the level of decision-making about conservation, the culture of indigenous communities is often underrepresented, or sometimes not even represented at all (cf. Anderson & Grove 1987, Einarsson 1993, Knight 2000a, Chatty & Colchester 2002). The latter is the case in Vassilikos where the indigenous culture in respect of the natural world has never been communicated outside the community itself. Conservationists, politicians and journalists in Greece have failed to identify anything worth mentioning in the relationship of rural Greeks to their animate and inanimate environment and this attitude has been particularly conspicuous in the environmental politics of Zakynthos. Thus, one of my objectives in this book, though not the primary one, is to explain the cultural background that informs the local resistance to environmental conservation.

Drawing upon a long anthropological tradition that emphasises long-term fieldwork and qualitative research I have chosen to elevate the human-environmental relationship in Vassilikos to the central theme of this book. The detailed ethnography presented demonstrates that *Vassilikiots*, the inhabitants of Vassilikos – whose views do not usually reach the world outside their community – do have a culturally informed approach towards the natural world. In the chapters that follow, I will trace it in their working, practical relationship with the productive resources of the land, agriculture, animal husbandry, and tourism. This working, pragmatic engagement of the Vassilikiots with their environment will serve as the common thread that unites my exploration of several other themes. These relate to the cultural features of landholding and cultivation, the working engagement of Vassilikiot men and women with tourism and farming, their attitudes towards domestic and wild animals, the classification of non-human living beings in their religious cosmology, and the passionate involvement of the local hunters with hunting.

The resistance of Vassilikiots towards environmental conservation provided me, not only with the initial impetus, but also with the ideal context for unravelling their human environmental relationship. Their desire to discuss their views about the natural world was enhanced by their bitter confrontation with environmentalists who campaign for the protection of rare species of animals, such as the loggerhead sea turtle,[1] and the establishment of a marine national park incorporating areas of the local coastal environment. Four communities on the island, and Vassilikos in particular, are affected by the conservation measures of this park, which mean that a number of the local inhabitants will be restricted in building for and developing tourism on their own land. Thus, those Vassilikiots affected by conservation, supported by their relatives and neighbours, vigorously protest against the establishment of the national park and the restrictions that emanate from it. Alongside the environmentalists' practices and ideals they set their own culture of relating to the land, cultivation, wild and domestic animals, stressing their own 'household-focused' priorities in their relationship to the immediate environment, understood by them as the field of daily work, toil and constant, hard labour.

The environmentalists, on the other hand, with their protective attitude towards what they perceive as the wild or untouched parts of the local environment, have introduced into Zakynthian environmental politics a polarisation between *nature* and *culture*, according to which 'nature is at its most valuable when it is untouched by human hand' (Milton 1996: 124). The culturally specific and essentialist character of this dichotomy (between the wild part of the environment and the social world) has been systematically underlined by anthropologists writing about nature and the environment in the last twenty years (MacCormack 1980; Strathern 1980; Ellen 1986a, 1996a, 1996b; Croll & Parkin 1992b; Milton 1993b, 1996, 2000; Ingold 1988, 1996; Descola & Palsson 1996; Green 1997; Berglund 1998). Following their insights, I have taken special care to present the Vassilikiots' worldview[2] about the natural world in terms of the particular cultural context of their day-to-day, practical interaction with the environment. In this respect, my analysis has been implicitly informed by Bourdieu's theory of practice (1977, 1990) and in particular by his emphasis on the cultural meaningfulness embodied in repetitive enactment of work.

For the most part, the ethnographic analysis that follows rests upon the foundations afforded by the vibrant and expanding body of anthropological literature on Greek (and Cypriot) culture. This includes a first wave of ethnographies that sketched broad portraits of social life in particular communities (Friedl 1962, Campbell 1964, Du Boulay 1974, Loizos 1975, Herzfeld 1985a), and a second wave of monographs that dealt with more specialised problems or themes (Loizos 1981; Danforth 1982, 1989; Herzfeld 1987, 1991a, 1992; Hirschon 1989; Cowan 1990; Stewart 1991; Seremetakis 1991; Hart 1992; Sant Cassia & Bada 1992; Faubion 1993; Dubisch 1995a; Panourgia 1995; Argyrou 1996; Karakasidou 1997; Sutton 1998; Just 2000; Kenna 2001a, 2001b). Paradoxically, such is the value of the ethnographic approach, that even these works that focused on highly specific topics, did not fail to address wider processes (cf. Just 2000: 20–8). Thus, although this book is about some people's relationship with the natural world, its various chapters directly or indirectly touch upon other aspects of their lives. They are my friends and 'respondents' in Vassilikos, men and women whom I call the Vassilikiots. In a very instrumental sense they all acted during my fieldwork as perceptive and sophisticated theorists, 'native anthropologists' to use an expression of Sutton's (1994: 241, 255; 1998: 36, 58), teachers, authors and critics of the ethnography that follows.

Introducing the self and the environmentalists

I will start with a revelation. It concerns an aspect of my engagement with the theme of this book, that though I have not kept it secret, I have rarely referred to. There was a time in my life when I used to declare myself an 'ecologist'. This term, divorced from its strictly academic meaning, is used in modern Greece to refer to 'environmentalists', people who are involved with the protection of the environment and share 'a concern that the environment should be protected, particularly from the harmful effects of human activities' (Milton 1996: 27). As such an aspiring 'ecologist' or environmentalist I used to contribute a large amount of my free time to the protection of the sea turtles and other rare species of animals threatened with extinction. It was during this period of my life that I visited Zakynthos for the first time in the 1980s, participating as a volunteer in the programme for the protection and scientific study of the sea turtle organised by a Greek environmental NGO, the Sea Turtle Protection Society (STPS). By that time, this particular reptile species had acquired 'neo-totemistic' qualities in my imagination (Willis 1990: 6): its fate held an elevated position among my own life-priorities.

I remember that in the narratives of my fellow environmentalists the relatively isolated turtle beaches of Vassilikos figured as a mythic place filled with stories of pioneering performances of environmental conservation. Not only were those beaches famous for the great density of egg-laying turtles, but also for the legendary perseverance of the 'turtle-protectors' (*khelonadhes*) in the face of the unwelcoming reception of the local population. Some of my colleagues and I perceived the build-up of tension between the indigenous community and the conservationists, not merely as a liability for the future of turtle conservation, but also as a battleground for establishing personal reputations. In this respect, the indigenous resistance to environmental conservation was treated merely as an adversity, similar to the inaccessibility of some turtle beaches or the lack of running water in the environmentalists' camp, an indicator of our faith in and endurance in the service of environmental ideals.

It must be noted here that in the early 1980s the environmental movement in Greece was still in an early stage. The newly established environmental NGOs[3] recruited graduates from biology and other hard sciences with minimal training in the social aspects of conservation. The majority of those recruits were predominantly, but not exclusively, middle-class urbanites (cf. Cotgrove 1982: 19, 34, 52, 93; Lowe & Goyder 1983: 10–11; Harries-Jones 1993: 46; Argyrou 1997: 160–64; Berglund 1998: 37) in their late twenties, inexperienced with, or ill advised about, life in the countryside and rural lifestyles in particular. Most of them espoused – at least in theory – a radical version of environmentalism[4] but, since their activities were focused on 'policy and practice', ideological disagreements regarding abstract environmentalist thought were a rare phenomenon (Milton 1996: 78; see also Norton 1991). More importantly, lack of long-term contact with the people living in the ecologically sensitive areas contributed to an unproblematic dismissal of the indigenous relationship with the natural world. As a young student in the social sciences at the time, surrounded for the most part by biologists, some of whom I much admired, I began to suspect that something was seriously wrong with the relationship between the 'turtle protectors' and the indigenous community.

My disillusionment with environmentalism started one Zakynthian summer night in the mists of the turtle egg-laying season. As a veteran 'turtle-protector' I was leading a team of four volunteers in taking a measurement in the conservation area. We all knew that the 'locals' (*oi dopioi*) – the environmentalists' generalising term for the indigenous people – were very agitated. For weeks they had been hampering my team's shifts by constant verbal provocation and threats of physical violence. Well prepared for trouble, we arrived at the centre of the local resistance armed with scientific equipment and a wireless transmitter tuned to the frequency of our headquarters. We were confronted by a growing crowd of 'locals' determined to stop us from entering the turtle beach. They were all men, exhausted from waiting for our late arrival after a long day's labour in the fields or in the local tourist establishments. They started shouting angrily, waving their hands and moving their sweating bodies in an impressive array of masculine postures. Struck by their passion, I put aside for a brief moment my environmentalist identity, to simply admire the unravelling of the indigenous performances, influenced, I guess, by Herzfeld's (1985a) – then newly published – *The Poetics of Manhood*. It was at this moment of perplexity that I was forcefully struck on the head – literally this time – by the wooden stick of one of the local protesters.[5]

When I recovered my senses I was already an eco-hero fleeing Vassilikos in a Land Rover sporting the insignia of WWF and the STPS. I spent the next few weeks secluded in a beautiful villa, belonging to a Zakynthian environmentalist, under strict orders not to appear in public. The local and national newspapers had already declared me in a critical condition and my superficial wound could not justify the magnitude of the publicity. But while my colleagues capitalised on the occasion to champion the cause of environmental conservation against 'the amoral and violent' enemies of the turtle, I was left alone, with plenty of time to contemplate my predicament in my beautiful Mediterranean prison. What I could not understand then was why those 'locals' were so gravely frustrated with conservationists such as myself: individuals whom I visualised as prompted solely by noble and selfless intentions. How could those 'locals' be so insensitive to the fate of such a unique reptilian species and what morality justified their lack of consideration, I wondered.

Ironically, this book addresses the same painful queries I sought to unravel as a result of that summer night in the late 1980s. My first wound in Vassilikos, which literally embodied the frustration of the local inhabitants, was gradually transformed into an indelible mark of ethnographic motivation. Arriving in the field two years later I embarked on the difficult task of persuading the Vassilikiots of my good intentions. The latter, masters in the art of concealment, were constantly testing my own skill in revealing information they had already decided to grant me (cf. Kenna 1992a: 152; Friedl 1970: 205, 215). Considering how frequently anthropologists who have worked with Greek communities have been referred to by their respondents as spies (Friedl 1970: 214; Campbell 1992: 152; Loizos 1992: 171; Handman 1987: 31; Herzfeld 1991a: 47–50; Just 2000: 3), I had anticipated a similar designation – 'the ecologists' spy'. I soon realised that most Vassilikiots did not remember or did not appear particularly bothered about my previous involvement in conservation. All of them, however, suspected that I was biased in favour of 'ecology' – the indigenous term for environmentalism – on the grounds that I was both an urban dweller and a well educated one. Not surprisingly, it took me some time to achieve rapport.

At first I was entrusted with safe, descriptive information suitable for a 'folklorist', the local, spontaneous translation of the term ethnographer (cf. Argyrou 1996: 26, 28), which was my self-asserted – and positively instituted on a national scale (Herzfeld 1986) – profile in the field. Later came the inside information about politics and the more intimate reflections of the Vassilikiots on the local environmental dispute. This came only after I had proved my good intentions by spending the first winter in Vassilikos, unlike the urbanites who, as my friends in the field maintained, fled to their comfortable city apartments at the first sign of bad weather. It might also have been the case that an act of 'complicity' on my part – a term understood by Marcus (1998: 108) as 'an integral but underplayed dimension of rapport' – helped me overcome the last barriers to the Vassilikiots' trust. On one particular occasion I helped some members of the community compose a written petition complaining against inconsiderate measures taken by the environmentalists.

My persistent efforts to prove myself as a fieldworker – in what was for me 'a foreign place, exuding the insecurities of anything new and the excitement of the unknown' (Panourgia 1995: 7) – and my attempt to deal with new personal and social identities, became intricately entangled processes. Transcending my urban identity so as to understand the life priorities of people living in the countryside was closely paralleled by the deconstruction of my middle-class, environmentalist ideals. In both cases, I had to abandon old prejudices in order to embrace new cultural lifestyles, a process of transformation (Loizos 1992: 172) that was forcefully effected only after a period of eighteen months. Dubisch describes most accurately what I felt, when she notes that 'different selves are constructed in the fieldwork process, different from those we have come to know in the context of our own society. These selves have some elements which are familiar to us, and some which are not' (1995b: 45). Awareness of my partial control over any one of these multiple selves grew out of the interplay of a multifaceted positionality (Bakalaki 1997: 518). I was an ex-environmentalist and an urbanite conducting fieldwork among, and feeling obliged to protect, people who felt harmed by environmentalists and other urban dwellers.

6

To renounce both my urban and environmentalist identity in the eyes of my new friends in the field, but also to get first-hand experience of the indigenous relationship with the land, I volunteered to participate on a daily basis in manual labour on the local farms. In this respect, the 'extremely physical nature' (Kenna 1992a: 155) of my fieldwork served simultaneously as a fruitful methodological approach and as an act of penitence for my cerebral pretentiousness. Covered with the distinctive grease of a day's work on the olive harvest, I would proudly walk along the village's main road, simply to induce Vassilikiots to comment that I was involved in 'real', manual work. My 'bodywork' (Kenna 1992a: 155) in this case embodied my claim to enter the 'back stage' areas of the local society, those regions, which according to MacCannell, are closed to outsiders (1976: 91–100).

Bakalaki (1997: 512) has recognised 'the estrangement from village life' experienced by middle-class Greek ethnographers, 'the children of people who at one time had to forget about their own villages', and who now perceive village life as exotic or – I would add judging from my own experience – uniquely authentic. Only a few years after my earlier pursuit of enduring experiences in the field of conservation, I found myself striving for the authentic in my performances as an ethnographer. My Vassilikiot respondents looked compassionately upon my inauthentic, over-motivated claims to the indigenous lifestyle. They gave me good lessons on how to slow down and accept the gifts of everyday life with patience. It took me several years to realise the importance of those relaxed, timeless afternoons in their company. It is those instances of rest from action that I now consider as my most fruitful and revealing moments in Vassilikos.

While I was confronting my own personal transformations in the field, my old companions in turtle conservation had not remained unaffected by the passage of time. On the contrary, those among them who continued to be involved with the protection of the environment became increasingly professionalised. Their adherence to the new ideological paradigm of environmentalism was becoming progressively more systematically articulated (Milton 1996: 77) and the NGOs that employed them became progressively institutionalised (Princen & Finger 1994: 8; for the Greek context, see Botetzagias 2001). Some completed postgraduate degrees in conservation in the UK, others learned from experience that a good relationship with the host communities was a fundamental prerequisite for the success of particular conservation schemes. According to my observations, however, this second wave of better-qualified environmentalists merely succeeded in addressing the social dimension of the existing environmental dispute in Zakynthos, without sincerely attempting to resolve the dispute itself. For example, during a prolonged winter stay on the island, one particular conservationist established firm foundations of trust and communication with one fishing community, only to find himself accused by his colleagues and friends of 'going native' (or 'local'). He was subsequently forced to resign his research contract.

Since the early nineties, in addition to STPS (the Sea Turtle Protection Society), two international environmental organisations, WWF and Greenpeace, have entered the scene of Zakynthian environmental politics. Both organisations established their own branches in Athens and recruited young Greek conservationists with previous field experience, including some veteran members of STPS. Thanks to this recruitment policy, the newly formed 'WWF-Greece' and 'Greenpeace-Greece' appeared remarkably well co-ordinated with respect to turtle conservation in Zakynthos. Less well co-ordinated *vis-á-vis* STPS, was another Athens-based NGO, the 'Mediterranean Association to Save the Sea Turtles' (MEDASET) (cf. Botetzagias 2001) which nonetheless demonstrated equal determination to rally to the defence of the turtle. Finally, a group of Zakynthian individuals devoted to the protection of the environment – mostly members of the island's educated and professional elite (cf. Argyrou 1997: 164) – formed the Ecological Movement of Zakynthos (ZOK). Its members provided all subsequent waves of outsider-environmentalists with valuable advice, and a firm foothold in Zakynthos.

But from the point of view of Vassilikiots and other Zakynthian commentators the increasing professionalism and co-ordination of the environmental groups was reached too late in the day to bridge the rift of misunderstanding and distrust already created between environmentalists and the local population. For the great majority of the Vassilikiots, all types of conservationists or 'ecologists' – whether socially enlightened or not – had already been firmly classified within one generalised category of hostile, untrustworthy, alien individuals. As Argyrou has argued with respect to anti-environmental sentiment in Cyprus, 'villagers and the urban proletariat, regardless of gender, consider environmentalism as the product of a foreign, morally inferior culture' (1997: 160).

To transcend the suspicion and resentment of an indigenous community in the context of an environmental dispute is not an impossible task. It requires time, patience and, primarily, willingness to examine the indigenous worldview as something valuable in its own right, a culturally informed alternative approach to relating to the environment. Exercising those virtues might appear unrealistic to non-anthropologists, such as the conservationists who have worked in Zakynthos. They will rightly complain that they lack the appropriate methodological and epistemological training, but also the time, to indulge in qualitative social research. This is why I am not attempting to reproach them for what they failed to accomplish, but rather to address them as one additional audience and share with them those indigenous views their policies have instigated. As I have clearly implied, environmental conservationists in Zakynthos do learn from their mistakes and have already become sensitive – to one degree or another – to the social parameters of their work. This book is a contribution to their efforts to become more considerate conservationists.

Chapter summaries

In the chapters that follow, I trace the interaction of Vassilikiot men and women with their immediate environment in a variety of different sets of meaningful and purposeful activity. Six separate themes are discussed in detail, providing an elaborate account of the cultural depth and richness pertaining to the indigenous human-environmental relationship. In the last chapter, I unite several issues raised throughout the book in a concluding analysis of the Vassilikiot human-environmental relationship, addressing, for a final time, the cultural dimension of the indigenous resistance to environmental conservation. My choice of presenting the most theoretically informed discussion of the monograph at the end, is intended to allow my more general conclusions to emerge from the thematic and comparative discussions offered in the six chapters that comprise the ethnographic body of the book (Chapters Three to Eight). This is also designed to facilitate reading for the less ethnographically inclined – for example, those involved in environmental policy and conservation – who will find a more concise account of the Vassilikiots' worldview towards the natural world in the last chapter.

In Chapter Two, I introduce the community which is the focus of this study, its social history and local environmental politics. I start with my first acquaintance with the community and my gradual appreciation of the Vassilikiot environment as a socially inhabited place. I continue with a concise account of the island's history, emphasising those social conditions from the past that are most intimately related to the present. Finally, I present a brief account of turtle-conservation in Vassilikos, focusing on the views of local inhabitants concerning the turtles, the environmentalists, the bureaucratic impasse created by the formalities of the state, and the long delay in the implementation of conservation legislation.

Then I proceed to systematically explore the multiple sets of values ascribed to the land in Vassilikos, the subject matter of Chapter Three. I also discuss how Vassilikiots articulate their opposition to the restrictions of environmental conservation. In the context of the environmental dispute, diverse narratives about the land are creatively stitched together to form a constantly transforming discourse, which reflects the unwillingness of the local landowners to be parted from their land. The third chapter also offers a short overview of the social and material circumstances faced by the inhabitants of Vassilikos when they were landless tenants (*semproi*) on the estates of wealthy landlords (*afentadhes*). Some of the older Vassilikiots describe in their own words their long and painstaking efforts to gain access to, and ownership of land and to escape from unfavourable tenancy agreements. The rich cultural significance Vassilikiots attribute to their land is related to those painful experiences in the past and their direct engagement with the productive resources of the land (tourism and farming) in the present.

The Vassilikiots' contention that both farming and tourist jobs entail a 'struggle' is carefully considered in Chapter Four. Their perception of work as a struggle testifies to a more combative attitude in general towards the natural world, which is informed by constant practical interaction with it. Success in both tourism and farming in Vassilikos involves constant care of the land and the tourist facilities erected on it, the continual struggle of the local farmers and tourist entrepreneurs to manage the disorderly and unpredictable elements of their immediate environment. Despite some contradictory pessimistic views expressed by some Vassilikiots in the context of their disappointment with either tourism or farming, the ethnographic evidence suggests that agricultural work and tourist enterprises in Vassilikos are less antagonistic than is usually thought. Vassilikiots frequently use resources and produce derived from their farms to sustain their tourist-related enterprises and vice versa. They interpret work undertaken in both economies as an investment in their households' wellbeing and self-sufficiency.

Similarly, household economic priorities and a spirit of co-operation between household members permeate the Vassilikiots' labour investment in the olive harvest, the topic examined in Chapter Five. During the harvest, Vassilikiot men and women work in small groups and divide their tasks according to a clear gendered division of labour, which is part of an olive cultivation culture with a long history on their island. Drawing primarily upon Strathern's (1988) work in *The Gender of the Gift* I present the Vassilikiot women's commitment to this work in terms of their desire to invest in meaningful household relationships. I also take the opportunity to examine the gendered dimension of agricultural work and its meaning as this is reflected by the local women. Their involvement in hard manual labour during the harvest is a further aspect of the Vassilikiots' pragmatic relationship with the productive resources of their environment.

The labour Vassilikiots invest in caring for animals is discussed in Chapter Six, which is a detailed ethnography of the human-animal relationship in Vassilikos. I explore the ways Vassilikiot farmers care for their animals, the ways they punish or complain about them, the repetitive, simple but exhausting tasks of their everyday interaction with them. In this respect, the systematic information presented in this chapter constitutes one of the rare cases in the anthropological literature where the human–animal relationship is studied for its own sake. Vassilikiots, like most other rural Greeks, maintain that they 'like animals, because animals are useful', but 'usefulness' in this context, as du Boulay (1974: 86) has accurately noticed, is not their 'sheer utility' but a necessary qualification for membership in the rural household; according to this perspective even human members are expected to be 'useful'. Vassilikiot farmers maintain that farm animals receive their 'care' (*frontidha*) and are expected, in turn, to reciprocate by respecting the 'order' (*taxi*) of the farm. 'Order' in the farm environment is defined and maintained by the farmers themselves with constant effort or 'struggle' and is directly related to the organisation of the household as an autonomous self-sufficient unit. The punishment and usefulness of particular animals, as well as the farmer-animal relationship as a whole, can be better understood when placed in this context of 'care' and 'order', which covers all domestic animals.

Wild animals, which play no part in this context of care and order, are treated by Vassilikiots with hostility (if they are perceived as harmful) or indifference (if their existence does not directly concern the lives and priorities of the local human protagonists). Chapter Seven, presents examples of the rare instances where Vassilikiot farmers discuss wild animals, and portrays their 'sorrow' in cases when wild animals prey upon domestic ones. Vassilikiots think of non-domesticated animals in terms of their own established presence in the local environment and their position as guardians of wellbeing and order on their farms. Their perceived authority over non-human living being of all kinds is axiomatic and can be fairly described as anthropocentric. It is underpinned by an elaborate religious cosmology, which emphasises God-given human 'dominion' over the natural world. To illustrate this issue, the chapter devotes some considerable attention to the classificatory insights of St Basil the Great, one of the most influential theologians of the Greek Orthodox patristic tradition. St Basil's taxonomic clarifications are presented as a coherent discourse, subjected – like ethnographic data – to anthropological enquiry and compared to my respondents' evaluations in Vassilikos.

Another category of wild animals, game that can be hunted, excites the imagination of the Vassilikiot hunters. Chapter Eight examines Vassilikiot men's intense engagement with hunting, which is considered in Zakynthos as a celebrated 'passion' (*pathos*). The ethnography presented illustrates the Vassilikiots' preoccupation with hunting both in the past and the present, describing how the local hunters boast of their guns and their skill in shooting. The opposition of the local hunters to hunting restrictions and the 'ecological' discourse is examined as further evidence of Vassilikiot resistance to environmental conservation. The chapter also considers male bonding and identity as realised in hunting performances and narratives. Hunting is treated as an all-male context, as in Papataxiarchis's (1988, 1991) description of the coffee-house, but one more positively attuned to the internal concerns of the rural household and thus more positively received by women.

In the final chapter, I unite several themes raised throughout the book in a conclusion that examines the human-environmental relationship in Vassilikos. I emphasise the confrontational spirit of the Vassilikiots' interaction with the environment and the meaningfulness of this combative or agonistic disposition for the local protagonists. Their engagement with the productive resources of the land is perceived locally as an investment in the wellbeing of individual households, which is dependent upon the close co-operation of family members in the fields of tourism and agriculture. Similar anthropocentric or household-centred priorities inform the work invested in caring for animals. Pursuing this observation, I devote some attention to the writings of anthropologists and social historians on issues related to human attitudes towards animals and anthropocentrism. I argue that attempts to classify indigenous attitudes to animals in generalising categories that stress the anthropocentric (or other) qualities of the human animal relationship are misleading. Vassilikiots' interaction with their animals and the environment resonates with complex cultural themes that combine the imposition of a humanly defined sense of order with an emphasis on care enacted through continual work investment in the environment itself. Vassilikiots maintain that they care a lot about their environment. As the ethnography that follows demonstrates, their caring practices reflect a rich, culturally embedded human-environmental perspective that deserves to be studied in its full complexity in the particular ethnographic context within which it is realised.

NOTES

1 Loggerhead Sea Turtle, *Caretta caretta*.
2 My use of the term 'worldview' is inspired by Pina-Cabral (1986) who chose it as a less inclusive alternative to 'culture' and a less codified synonym of 'cosmology' (1986: 4–5). He traces the term back to Daryl Forde and his *African Worldviews* (1954).
3 Princen & Finger define environmental NGOs as 'those non-profit groups whose primary mission is to reverse environmental degradation or promote sustainable forms of development, not pursue the objectives of governmental or corporate actors' (1994: 16). Harries-Jones's portrayal of environmental NGOs emphasises their involvement in social advocacy – which he defines as the 'positive form of protest carried out by a defined network or group' – and their non-party, extra-parliamentary, non-violent character (1993: 44).
4 Milton (1996: 74–8, 205–6) carefully outlines the general tendency of several social scientists who study environmentalism to differentiate between more radical and more conservative versions of environmentalist attitudes and thought (cf. Cotgrove 1982, Hays 1987, Norton 1991, O'Riordan 1976, Worster 1977). It must be noted, however, that Norton questions 'the continuing usefulness of categorising environmentalists as two exclusive groups' since he observes the co-operation of their environmental activities in practice (Norton 1991: 9–10; Milton 1996:77).
5 Kemf (1993) provides us with a short description of a similar incident in Kalamaki, a small community not far from Vassilikos. In the summer of 1992, Michalis Antypas, a veteran environmentalist working for STPS, was injured 'while distributing information leaflets to tourists' by a group of Zakynthians affected by conservation (ibid: 187).

2
VASSILIKOS: PAST, PRESENT AND TURTLE TROUBLES

Getting to know Vassilikos

Vassilikos is the name of a narrow peninsula, in the southwest corner of Zakynthos.[1] It is also the name of a socially inhabited space, the local administrative community. Its three major settlement concentrations hardly fit the stereotypical notion of a Greek village as a tight cluster of households with a central square and a church. The community itself is composed of households dispersed in among green fields and olive groves. As one of the currently available tourist guides maintains: 'Vassilikos is more of a concept than a place. The houses are spread over a considerable area, in among the greenery; the fields and orchards are watered by the abundant streams and there are many good beaches' (Toubis 1991: 92). Mount Skopos is the backbone of the Vassilikos peninsula. From its summit down towards the plains of Vassilikos, the habitable strip of land between the sea and the mountain becomes narrower and more fertile. This is Vassilikos proper, but the mountainous region of Skopos and the area called Xirokastello adjacent to it, are part of the 'community of Vassilikos' (*koinotita tou Vassilikou*), and the people living in the area identify themselves as people of the same community.

When outsiders talk about Vassilikos they almost always use the terms 'beautiful' and 'nature'. Greek visitors often put the two words together to form the expression '*oraia fysi*' (beautiful nature), with which they reveal their aesthetic appreciation of the local physical environment. In the past, the middle or upper classes from the island's capital, who used to frequent Vassilikos for recreational purposes (a family jaunt or a hunting excursion), used to describe Vassilikos as 'beautiful countryside' (*omorfi exokhi*). Nowadays thousands of Greek and foreign tourists spend their summer holidays in Vassilikos and declare themselves fascinated by its 'nature' and 'beautiful beaches'. Likewise, those environmentalists who have visited Vassilikos on their campaigns do not hesitate to express their reverence for, and determination to protect, the 'beauty' of the 'natural' ecosystem.

'Dream of getting-away and relief for the sad man, is the enchanted Vassilikos' writes a Zakynthian scholar (Konomos 1979) in a literary portrait of the land's natural beauty. His perception reflects 'an outsider's view of beauty and tranquillity' (Waldren 1996: xv, 39). This was the view of the town dwellers of the island's capital, the urban Zakynthian elite who traditionally regarded Vassilikos as the countryside, the place to visit on May Day for a picnic close to nature. Other outsiders have chosen to reflect on other features of Vassilikos' physical environment. A visiting hunter, for example, will emphasise the vicinity's ideal position as a spot for hunting turtledoves, while a conservationist will almost certainly stress the importance of the local beaches for the reproductive cycle of the loggerhead sea turtles. A tourist, of course, will praise the same beaches for their fine quality of sand and their warmth, natural properties that have also attracted the turtles to lay their eggs there. All these diverse categories of visitors have been specifically interested in Vassilikos as a hunting ground, a picnic site, a tourist resort, or a natural ecosystem, but rarely as a place where people live.

My first task after arriving in the field as an anthropologist was to fill this empty natural landscape with human voices, memories and narratives. I soon realised that getting to know Vassilikos from its inhabitants' point of view required time. As one of my first local respondents rhetorically explained, 'you have to live and work on this land to feel it'. But I had not yet worked the land of Vassilikos, and although I had visited the place before, either as a conservationist or as a tourist, I could not legitimately claim that I had 'lived' it. To testify to my own initial perceptions of 'naturalness' in a highly 'worked-upon' and 'lived-in' landscape – in terms borrowed from Barbara Bender (1993b: 1–7) – I will present three extracts from my fieldnotes, describing the same location during three different stages of my fieldwork. I take my cue from Bender (1993b: 3–10), who refers to a novel by V.S. Naipaul, *The Enigma of Arrival*, and the progressive discovery by the Trinidadian-Indian author of the English landscape. Bender quotes Naipaul:

> ...when I grew to see the wild roses and hawthorn on my walk, I didn't see the wind-break they grew beside as a sign of the big landowners who had left their mark on the solitude... I didn't think of the landowners... I thought of those single-petalled roses and sweet smelling blossom... as wild and natural growths (Naipaul 1987: 24; quoted in Bender 1993b: 7)

These sentences looked only too similar to some notes in my fieldwork diary. Like Naipaul, each time I took the same walk I saw a different land, a more meaningful land. This is what I wrote in my fieldnotes about a property hidden from the gaze of the passing traveller:

> Today I was walking on the land of the 'big landlord' (*tou megalou afenti*). 'It is all his land', I realised. I noticed a big structure, like a mansion, but I couldn't clearly see the buildings behind the tall white wall. The place looked uninhabited although not deserted. Everything was clean and orderly. I was told that the landlord and his family traditionally live on another property closer to the town. I was also attracted by the deserted buildings all around. One of them is a deserted olive press. The others are small squat houses made of brick. Most are completely ruined, but two of them are renovated and have been transformed into beautiful cottages like those rented to tourists. I noticed the row of huge trees around the mansion, mostly eucalyptus. I enjoyed walking the path parallel to the trees with its beautiful view. 'Time has added an element of mystery and aesthetic beauty to those ruins', I said to myself, gathering an old rusty tin of sugar from the ground...

As already becomes apparent, my first views of Vassilikos were saturated with the semi-romantic aesthetic appreciation of the town dweller. Some information about the actual place was blended with beautiful images and landscapes, giving the point of view of the outsider. A few months later, being more intimately attuned to the life of the community, I noted:

> Considering the main road in Vassilikos is the artery of the village's social life, the old mansion is located some distance from the road, yet not that far away. This means that it can be overlooked by the tourists and visitors. I could imagine, though, that in the past this was the centre of social and economic life. Considering the scattered pattern of settlement in Vassilikos, the area around the mansion would have been populated by many peasant workers in the past, poor people living in small dank cottages. The landlord's mansion would have been the focus of activity, or even the locus for managing village resources.

I crossed the same area for a third time, a year after my arrival in Vassilikos. This time I was not merely contemplating the features of the landscape, but I was trying to help an old Vassilikiot man with shepherding his sheep. I recorded:

> While we were herding his flock across the landlord's land, Old Dionysis pointed to the landlord's mansion (*arkhontiko*). He talked about the warehouses, barns, the animals (*ta zontana*), the carts and coaches (*kara kai karotses gia anthropous*), the 'many horses'. 'There used to be several hamlets around this mansion', Old-Dionysis said and pointed to the ruined, small houses I had noticed before: 'There the landlord used to organise workers from other villages and his own *semproi* (landless tenants). He had fifty families of *semproi* living on his land'!

The social dimension of life in Vassilikos is not visible at first sight. From one point of view, it does not come as a surprise that some visitors, tourists or conservation-

Figure 2 Vassilikos road sign (misspelled!...)

ists, fail to register the local environment as a social place. After all, as my Vassi-likiot friend maintains: 'you have to live and work on this land to feel it'.

Social history

The small peninsula of Vassilikos has been inhabited since antiquity. Archaeolo-gists have identified the remnants of Neolithic and Mycenean settlements and artefacts from later periods (Sordinas 1993, Kourtesi-Philipaki 1993, Kalligas 1993). Homer, a less reliable source, maintains that Zakynthians, as subjects of the state of Ithaca, participated in Odysseus's campaign against Troy and flirted with Penelope as potential suitors. Ancient Greek mythology also portrays Artemis, the goddess of hunting and the wild, enjoying wandering in the woods of Zakynthos, and there is evidence that she was honoured and venerated by ancient Zakynthians, much as modern Zakynthians nourish a great love and pas-sion for hunting.

During historical times, Zakynthians as citizens of an independent city state were involved in the Peloponnesesian wars in the fifth century BC, helping

Kerkyra (Corfu) and Athens in their campaigns against Corinth and Sicily respectively (Thucydides I, 47; II, 66; Kalligas 1993; see also Sidirokastriti 1993, Toubis 1991). Later the island was ruled by Macedonians and Romans, and during the late Roman period it was subjected to endless incursions by 'barbarian' hordes and pirates. Visigoths, Huns, Vandals, Saracens and Normans destroyed whatever was left to be destroyed on plundering expeditions to the western borders of the Byzantine empire. During the thirteenth and fourteenth centuries the island was controlled by two Frankish families, the Orsini (1185–1375) and the de Tocci (1375–1479), subjects of the Kings of Naples. Their allegedly inefficient rule was followed by an Ottoman invasion (1479) which devastated the remaining population and its material resources (Konomos 1981).

Soon after the Ottoman raid, the Venetians, who observed the dramatic events of 1479, negotiated with the Turks for the proprietorship of the island. For the Venetians, control of Zakynthos was an objective they had planned carefully some years before 1485, the year their official rule commenced (Konomos 1981: 19–25). But the Venetians found the island in a state of complete desolation. Most of the lands were deserted and the once cultivated fields covered with wild vegetation due to neglect. The Venetians immediately issued proclamations to all neighbouring Venetian provinces in mainland Greece welcoming new settlers on the island (ibid.: 27–38, 45). The island was consequently repopulated primarily by Greek soldiers serving the Venetian army, their families and other Greek-speaking subjects of the Venetian Republic who fled from Ottoman territories to enjoy the protection (much needed at the time due to widespread piracy) and moderate suzerainty of Venice.

The years of piracy, plunder and relative depopulation preceding and following the establishment Venetian rule are depicted in the collective memories of the present-day Vassilikiots as 'the time when the land was deserted'.[2] Two such accounts are narrated below:

> The island was deserted (*erimo*). Two families came from Peloponnese, two families with sheep... They came to Zakynthos to escape Turkish rule. Then the Venetians issued a proclamation (*vgalan firmani*) and noblemen (*arkhontes*) came to settle on the island. Here in Vassilikos there was only a monastery, the monastery of Akrotiriotissas. The monastery used to take payments from Venice to save shipwrecked people (*tous pnigmenous*).

> Vassilikos was deserted. No one wanted to live here, because of the Saracens (*Sarakinoi*). Then one came...another one followed... This is why we have different names. It is not like Cephalonia, where everybody's name ends with '-atos'. You see, at this time it was not forbidden to cut trees (*logous*) and bushes (*thamnous*). If you could find deserted land you could settle on it...

Vassilikiots' consciousness regarding the island's history stretches back to the 'time of the Venetians'. They often point to the large olive trees (*dopies*) on their land and say: 'those trees are have been here since the old days, the time of the Venetians!' These trees are planted in rows equidistant from each other, a plan that reveals, according to my Vassilikiot respondents, a Venetian practice. The older men also persist in referring to one particular placename, '*Tis Martas t' aulaki*' [the trench of Malta][3], explaining that:

> There used to be a long trench here. In the old days the Venetians were trying to make a passage (*perasma*) to avoid the cape of Gerakas [the end of the Vassilikos peninsula]. They wanted to sail their ships through it. They dug and dug, but they never managed it (*dhen ta kataferan*).

The Venetians retained their control over Zakynthos for three hundred years (1485–1797). During that period, the capital of the island expanded out of its fortified medieval settlement, the population increased, architecture and commerce flourished. Wealth and prosperity, however, were the privilege of an elite: the *nobiloi*, a tough feudal aristocracy, emerged as the dominant class of Zakynthian society. Its members were recorded in the *Libro d'Oro*, the Golden Book. In Zakynthos, unlike other Venetian territories such as Cephalonia, membership of the *Libro d'Oro* was strictly limited to approximately 374 members (Zois 1963; Roma 1967: 478). This, as a consequence, excluded the growing urban middle class from the benefits of various political and economic privileges, and culminated in social unrest. The most wealthy merchants and artisans of the capital encouraged the poor urban dwellers of the island's capital – who were scornfully referred to as the *popolaroi* (common people) by the aristocracy – to rise in rebellion. This became known as the 'rebellion of the Popolaroi' (1628–31).[4] The rebellion ended with a victory for the aristocrats, who further secured their privileged status, and whose power remained unchallenged for the next three hundred years.

During the rebellion, the lowest stratum of Zakynthian society, the *semproi* (landless tenants), fought bravely on the side of their feudal masters, the same aristocrats who systematically exploited them (Konomos 1981: 62, 107; see also, Roma 1967: 443, 1971: 29–30). The *semproi* composed a large underclass of peasant labourers working on the Zakynthian feudal estates, who in the context of the Venetian era can be justifiably called 'serfs'. They were often recruited as soldiers to accompany their masters on Venetian military campaigns. The Zakynthian landlords, it is said in Vassilikos, 'used to have rights of life and death' over 'their' *semproi*:

> In the old days the master was the one to grant permission for a *sempros*'s marriage. The master was the one to sleep first with a *sempros*'s wife on the first night of the marriage. The master was the one to decide about everything.

Mylonas (1982), a Zakynthian scholar, describes that when the feudal right of 'taking the maidenhead' (*parthenofthoria*) was abolished (he does not exactly mention when), the semproi men, on the second day of their marriage, used to hang their trousers from a tree and shoot at them. 'That was the proof that the first night of marriage was theirs...', the same author maintains and concludes, 'this custom was practised in Zakynthos until our days' (Mylonas 1982: 86–7).

In the three hundred years following the 'Rebellion of the Popolaroi' the feudal aristocracy remained in power, and the landless tenants of the countryside continued to serve their feudal lords obediently. An exception to this situation were the inhabitants on the mountainous southern and western side of the island, who managed to escape feudal exploitation, retaining the status of 'free peasants' (*eleftheroi khorikoi*) (Roma 1967: 443–4; Konomos 1981: 87), especially in those cases where their land was not fertile enough to attract the interest of the aristocracy.[6] Some of those mountain-based Zakynthians, proud of their 'defiant independence' (Herzfeld 1985a: xii), still call the villagers of the plains and the people of Vassilikos, 'faithful-to-the-master serfs' (*afentopistous semprous*) [see, also Chapter Five].

In the meantime, the urban *popolaroi*, like the mountain villagers, retained their desire for self-determination and in the eighteenth century identified with the ideals of the French Revolution. When the French army arrived on Zakynthos, ending Venetian rule in 1797, the *popolaroi* celebrated with enthusiasm what they believed to be the end of an oppressive regime, and publicly burned the *Libro d'Oro* (Konomos 1981: 169–211). Their celebrations however, were premature, as the democratic French did not remain in power for long. After brief periods of Russian (1798–1800) and, later imperial French (1807–9) sovereignty, the island became a British protectorate and the power of the aristocrats was restored (Konomos 1983, 1985; Hannell 1989: 107–9, 124–6).[7]

It was only after 1864, when Zakynthos was incorporated into the newly founded Greek state that the power of the aristocracy was drastically limited. By the turn of the twentieth century, the Zakynthian middle class had gained a dominant position in local political and social life (Konomos 1986: 20). But while union with Greece enhanced the political and social position of the middle class, it led, at the same time, to a period of cultural and economic decline.[8] In the early part of the twentieth century commerce deteriorated and cultural activities gradually declined. The once renowned capital of Zakynthos, which had developed over centuries a distinctive Ionian cultural identity (cf. Herzfeld 1986: 24–30),[9] was gradually transformed during the second and third decade of the twentieth century into a mere provincial town.

Unlike the Zakynthian middle class, who successfully ended their centuries-long battle with the aristocracy, the landless tenants living in the island's countryside remained dependent on the landlords until as recently as the Second World War, and in some isolated areas like Vassilikos, until even later.[10/11] While novelists and local historians have devoted considerable attention to the struggle between the bourgeois and the aristocrats, the *semproi* or landless tenants of the countryside and the conditions they lived in has remained a topic of inquiry overlooked by Zakynthian scholars and writers – despite the sympathetic attention of some.[12] During my fieldwork I once visited an elderly Zakynthian woman, the wife of a prominent Zakynthian writer of aristocratic descent. When I tried to explain that I was studying the farmers of Vassilikos and their way of life, she looked at me with amazement and added: 'What will you find worth writing about there?'[13]

In fact, until as recently as the early 1960s, most inhabitants of Vassilikos were *semproi* working on the estates of landlords. One of the landlords was the descendant of an old aristocratic family with land rights to the area since Venetian times. He owned most of the land in Vassilikos and my respondents refer to him in their narratives as 'the big landlord' (*o megalos afentis*). The rest of the landlords were Zakynthians of bourgeois origin, living in the island's capital but owning landed property in Vassilikos. Until the early 1980s all these categories of landlords were addressed by their *semproi* as 'masters' (*afentes*); the big landlord often being called by his aristocratic title, the Count (*o kontes*). The *semproi* were entrusted by their landlords with parcels of land to cultivate, and were entitled in return to a small portion – usually approximately one fourth (*quarto*) – of the agricultural produce. The particular form of the rules managing the economic relationship between landlords and peasant labourers (*kopiastes*) were called *sempremata* in Vassilikos. As I will describe in detail in the following chapters, different modes of *sempremata* regulated different kinds of cultivation or animal husbandry. Undertaking an agricultural project according to a particular pattern of *sempremata* is called *sempria* in Zakynthos.

Even nowadays, some Vassilikiots continue to undertake *sempria* arrangements, but the greatest portion of the produce is now allocated to them. Almost all of them own small parcels of land and their dependence on landlords has decreased significantly. The descendants of the 'old-time' landlords – some of them still owning considerable areas of land – are still treated with respect by the majority of their ex-tenants and their families. Present-day Vassilikiots, however, make all major decisions concerning their lives and their economic ventures with total independence. Their freedom in choosing between a variety of possible occupations is enhanced by the recent rise of the tourism economy, which often enables the more entrepreneurial individuals to engage in more than one economic activity at the same time.

It is true that before the introduction of tourism in the late 1970s and early 1980s, the great majority of Vassilikiot farmers were seriously constrained by poverty and simultaneously demoralised by the social depreciation of farming lifestyles (cf. du Boulay 1974: 246–56). Tourism, however, was to become the panacea for the Vassilikiots' economic problems. It gave new impetus to the existing economic enterprises and gave rise to several new ones. To illustrate this, more than thirty *tavernas* or restaurants operate in Vassilikos during the summer, while the permanent population of the village does not exceed six hundred residents. Car rentals, renting out canoes and sun-umbrellas on the beach, mini-markets and, most importantly, room rentals – almost every household has some 'spare' rooms for rent – complete the catalogue of typical tourist enterprises in the vicinity.

What is more important, however, is that tourism did not make the pre-existing agricultural economy redundant. As I shall carefully illustrate in the course of this book [see, in particular, Chapter Four], Vassilikiots make more profit from tourism than from agriculture, yet they do not appear set on severing their working relationship with the land. They all invest their energy into making the resources of their land productive, and these resources include farming and animal husbandry, but also tourism. The newly built enterprises of tourism, little villas, apartments or restaurants, require work for their maintenance and the constant management of the local natural environment. Even since the old days, the times of the Venetians and the feudal landlords, Vassilikiots have been constantly involved in work, or as they prefer to call it, a 'struggle' (*agonas*). Unlike the past, they now enjoy the fruits of this constantly enacted effort and appear determined to stop outsiders – environmentalists or others – from infringing their justifiable claim to progress (*prokopi*).

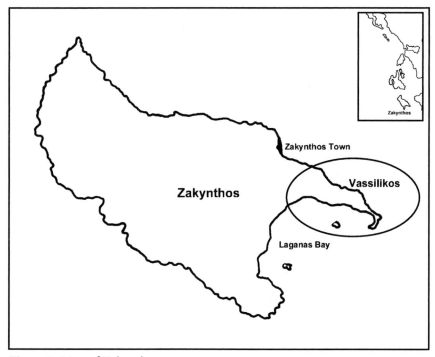

Figure 3 Map of Zakynthos

Turtle politics

In the 1980s the sweet smell of tourism development began to spread over the land of Vassilikos. Blended with the sea breeze and the distinctive aroma of the Zakynthian vegetation, the aspirations of local farmers for a better future were coming, progressively, closer and closer to realisation. There were new unexplored possibilities in the tourism job market, original entrepreneurial dreams to be realised, and the Vassilikiots, who had been, for most of their recent history, oriented towards the land, started looking with greater optimism towards the beach. Those who had considered themselves unlucky for inheriting property near the seashore – the land most unsuitable for cultivation – were now seen as the luckiest of all (cf. Herzfeld 1991a: 41, 73, 154; Boissevain 1996b: 10; Zarkia 1996: 150).

Vassilikos is not, and has never been, a fishing community. Before the introduction of tourism, Vassilikiots' engagement with the local marine environment was rare and opportunistic: just a handful of individuals with some expertise in fishing and two or three fishing boats. Despite their devotion to a farming lifestyle, however, most Vassilikiots – and especially the men – had one or two stories about fishing to tell. Their knowledge of marine fauna included detailed descriptions of edible fish and a few scattered references to other non-edible creatures: the dolphin, the seal and the turtle. Those three marine species were often blamed for the destruction of fishing nets. Despite this occasional damage, Vassilikiots, who were not fishermen by profession, admired the dolphin, detested the seals and were indifferent towards the turtles (see, also Chapter Seven). 'We never paid special attention to the turtle', my oldest respondents remember, 'the turtle was not harmful or useful to anyone, it didn't bother anyone!'

Unlike other people threatened by conservation who confront animals stigmatised as harmful or dangerous (cf. Knight 2000b, Richards 2000, Marvin 2000b, Lindquist 2000), Vassilikiots had no particular reason to regard or disregard the turtles. They simply 'couldn't ever imagine' these wild animals to be 'worthy of so much attention'. The local farmers' attention has always been devoted to their 'own' animals, the domestic creatures of the farm. As I will illustrate in the chapters that follow, the human-animal relationship in the environment of the farm is prescribed in Vassilikos by rules of reciprocity. The human caretakers offer their care and protection and the domestic animals their obedience and 'usefulness' – 'use' being understood here, not as utility in monetary terms, but as any form of direct or indirect contribution to the farming household of which domestic animals are legitimate members (du Boulay 1974).

The turtles, on the other hand, failed to correspond to any legitimate, locally ascribed usefulness. Their meat (and their eggs) were considered inedible and for a good reason: the meat of the Loggerhead turtle – unlike that of the green turtle – is too hard to cook, at least by the demanding standards of Zakynthian cooking. With respect to the second, more widely recognised, virtue of the Loggerhead turtle, its vital role in keeping the number of jelly-fish under control, not much mention was made of this in Vassilikos. Before the introduction of tourism, the landlubber Vassilikiots had little opportunity to listen to tourists complaining about the increasing numbers of jellyfish in the Zakynthian waters. The turtles, the jellyfish and the marine ecosystem as a system of interconnected organisms, were all concepts of no particular consequence for the lives of the local farmers. Their moral universe and their priorities were centred on the care of their animals, the protection of their cultivation and the future of their children.

When the 'protectors of the turtle' (*oi khelonadhes*) – the members of the Sea Turtle Society (STPS) – arrived on the island for the first time, Vassilikiots approached them with curiosity. It was quite surprising for them to see young, educated people caring so much about a wild animal 'like the turtle' (*san tin khelona*). At first they did not perceive any particular threat; they simply expected these strange researchers – whom they collectively referred to as 'the ecologists' – to finish their measurements and leave. But 'the ecologists' left the island briefly, only to return the following year. Since then they have visited the island every summer, more numerous and better organised each time. They attracted to their side the letter of the law, the powerful emotions of the general public and the support of internationally renowned environmental NGOs.

Soon after the first arrival of the environmentalists in Zakynthos, a Presidential Decree made the Loggerhead Sea Turtle a protected species in Greek waters (in 1981). Subsequent decrees introduced the first restrictions on tourism development and other human activities in the areas close to the turtle breeding grounds (in 1982, 1984, 1990). 'The original state of bewilderment' experienced by the Vassilikiot landowners at the imposition of conservation (Campbell 2000: 142), was succeeded by anxiety and exasperation. The Vassilikiots had, in fact, a very good reason to worry: three of the six turtle-hatching beaches in Zakynthos lie well within the confines of their community. What was to become of the neighbouring Vassilikiot properties and to what extent would the Vassilikiots and their tourist guests enjoy access to those beaches?

The news about the Presidential Decrees was followed by the openly advocated plans of the environmentalists to establish a Marine National Park in Zakynthos. Those Vassilikiots with properties adjacent to the turtle-hatching beaches were now seriously alarmed. In the beginning, they hoped that either the ban on building activity would be rescinded or satisfactory compensation would be paid by the government. But as the years went by, they realised that the state authorities were reluctant to offer any kind of significant monetary compensation for the conserved lands. Furthermore, the effective advocacy of the environmental groups brought about new environmental laws imposing further restrictions.

The five environmental NGOs[14] that have been involved with the protection of the sea turtles in Zakynthos, despite occasional disagreements or competition between them (Botetzagias 2001),[15] have been unanimous regarding one major objective: the creation of a National Marine Park in Zakynthos. The Mediterranean subgroup of the Loggerhead turtle[16] is threatened with extinction due to habitat loss and one of the most important factors for its survival is the protection of the few remaining hatching beaches in Zakynthos. For the egg-laying of the turtles to take place, the requirements are a minimum of noise and light pollution on the land surrounding the 'egg-laying beaches' and a virtually total lack of human presence on the beaches themselves (Margaritoulis et. al. 1991, Cape 1991, Arapis 1992). Thus, turtle conservation, according to the environmentalists' point of view, presupposes the imposition of serious restrictions on tourist development, or any kind of development, on these particular beaches and the surrounding land.

The exact description and parameters of these restrictions soon became a topic for debate. On a few rare occasions, Vassilikiots succeeded – by dint of considerable effort and resourcefulness – in influencing some of the representatives of the local authorities and some individual members of the parliament in their favour. Ministers and senior officials in Athens, however, under continual pressure from the press, environmental NGOs and the EU, were compelled to appear environment-friendly. Unprepared to pay the actual cost of environmental protection – that is, the due compensations to the affected landowners – they escaped responsibility by hiding behind the informal indifference of the bureaucracy (Herzfeld 1992). Their official response to any query regarding turtle conservation in Zakynthos was to make a formal presentation of plans laden with intricate details, restrictions, instructions and guidelines for the establishment of the Marine Park. Planning, though, was always under way, and when it was finished, more needed to be done.

At the abstract legislative level, the exact specifications concerning the formal constitution and administration of the Park had always remained a matter to be decided upon. In practice, however, the immediate restrictions of the Park were put into effect as early as the late eighties, owing to a series of short-lived laws and decrees. Initially, the affected landowners had been informed that they would not lose legal possession of their land. They were, in fact, allowed to exercise all forms of traditional farming on their 'conserved' property. During the early stages of the conservation restrictions, local politicians and state officials explained to them that the conservation measures were, in fact, temporary and that a more drastic solution to their problem was imminent. The reluctance of the state bureaucracy to pay the cost of conservation was thus officially rationalised by a rhetoric of incompleteness and temporality. Under the pretext that more environmental planning needed to be done and then assessed, delay and postponement in finalising the conservation measures was instituted as the official policy of the state. As Herzfeld has noted in a conservation dispute in Crete, where bureaucracy and a need for precise documentation is involved 'nothing is more permanent... than the temporary' (1991a: 251).

By the early 1990s, the time of my fieldwork, those Vassilikiots most directly affected by the conservation measures were already frustrated and demoralised.

Silenced and disempowered when confronted with bureaucrats and other state officials – like most other peripheral Greek actors (Campbell 1964; Loizos 1975; Herzfeld 1991a, 1992) – they re-directed their resistance at local targets. Considering the environmentalists responsible for the conservation prohibitions, they tried to hinder their movement and activities on their land. For a short period in the early 1990s, Vassilikos became an 'ecologist-free' zone, but its inhabitants soon realised that their resistance was effective only within the confines of the local community and its immediate vicinity.

In the meantime, the environmentalists responded to this challenge by displaying an ever-greater commitment to their objectives. They became better organised and initiated information programmes for the public and the tourists. Paradoxically, with their resistance to turtle conservation, the Vassilikiots succeeded in giving the environmentalists the publicity they so much required. Not surprisingly, the environmentalists won the battle of the media, secured funding support from the EU, and demonstrated that there was a good cause for them to justify their involvement in the Zakynthian environmental politics. They appear determined to continue their own battle to save the turtles, for as long the latter remain an endangered species, or at least, for as long the Vassilikiots continue resisting.

In the year 2000, when I finished writing this book, rumours were heard that a final agreement about the management of the National Park was about to be reached. However, anyone who has spent time reflecting upon the politics of turtle conservation in Zakynthos, knows very well by now that any approach towards the future of the Park that focuses narrowly on legal parameters and convoluted regulations, blindly complies with the politics of bureaucratic indifference and formality instituted by the state (Herzfeld 1992). As the foremost sociologist of modern Greece has pointed out, 'a striking characteristic of political and cultural practices in the Greek social formation is the extent to which conflicts and debates take formalistic-legalistic character, shifting the attention of the masses away from "substantive" issues' (Mouzelis 1978: 134). From the point of view of those Zakynthians affected by conservation, the excessive formalism of Greek legislation is just the pretext used by the bureaucrats to avoid paying the due attention to their demands for compensation.

The history of turtle conservation in Zakynthos clearly suggests that environmental conservation cannot be successfully implemented without the consensus of the local population. In those cases, however, where the local political and administrative structures are not prepared to carry out the burden of conservation policies, neither decisive legislation, nor even the provision of economic incentives for the affected parties, will necessarily guarantee the co-operation of the local community (cf. Richards 2000: 97). This is the case in Vassilikos, where a necessary prerequisite for ensuring the co-operation of the local inhabitants is a conservation policy that takes serious account of the indigenous relationship with the land and its productive resources. The chapters that follow will examine, step by step, this complex relationship.

I will end this chapter with the views of two Vassilikiot landowners affected by conservation:

> The people of the village are very angry about the ecologists. At the beginning a few of them came. We gave them hospitality. We welcomed them on our land. They said they were counting the turtles... Then they kept on coming. More and more of them, every summer. They said we couldn't build on our land. It all happened because of the turtle...

> We don't want the ecologists on our land. They did harm (*zimia*) to several people here. They try to tell us what to do with our property. What to do in our own fields. We didn't go to their place to tell them how to run their own homes. If the ecologists care for the turtles, then why don't they take them onto their own property?

NOTES

1   Zakynthos lies seventeen nautical miles west of the Peloponnese and fourteen south of Cephalonia. Its overall size covers 406 square kilometres and it has a population of approximately 400,000 inhabitants.
2   See also, Xenopoulos (1984: 54).
3   See also, Maria Sidirokastriti (1993).
4   Dionysios Roma (1971: 15), Maria Sidirokastriti (1993) and other Zakynthian scholars claim that the 'Rebellion of the Popolaroi' was the first urban middle-class revolution in the post-feudal Mediterranean.
5   Konomos (1981: 43) explains that the Venetian democracy did not favour land-based feudalism in theory, but made notable exceptions to this principle when this system was applied to its colonies.
6   Roma (1967: 437–445) treats the category of the 'free peasants' (*eleftheroi khorikoi*) as a separate class from the landless tenants (*semproi*), in the same way that he distingusihed the urban, lower-middle-class *popolaroi* from the rich members of the upper-middle class. According to this system of stratification the aristocrats or *nobiloi* comprised the fifth and highest Zakynthian social class. It is worth mentioning that although Roma's primary objective was to provide a long mytho-historiographical account of an imaginary aristocratic Vassilikiot family, his plot was closely based on the archives of the long-recorded history of his own family. Consequently his novels are supported by long endnotes which present historical facts, published and unpublished archival information, even, short dissertations on Zakynthian social history during the Venetian era.
7   In my brief account of Zakynthian social history I have deliberately chosen to stress the Venetian era, a formative period for the crystallisation of social stratification in Zakynthos. The British did not significantly challenge the pre-existing social divisions and the Zakynthian aristocracy retained a great deal of its former power until the unification of the Ionian Islands with Greece in 1864 (Konomos 1983, 1985). The British arrived in Zakynthos in 1809 and the island officially became a British protectorate in 1815 (Konomos 1983: 189; Hannell 1989: 105). Hannell identifies the same basic social distinction between the nobility, the middle class and the manual labourers, and testifies to the exploitation of the poorer strata by the aristocracy (Hannell 1989: 108–9).

8 This comment is indicative of my opinion, but also of several Zakynthians I have met in the field.

9 It is worth mentioning that Zakynthos is the birthplace of distinguished poets such as Dionysios Solomos, Ugo Foscolo and Andreas Kalvos, and several other renowned intellectuals.

10 Roma (1967: 445) observes that in Zakynthos the practice of unfavourable tenancy agreements persisted even after the nationally instituted expropriation of large land estates in mainland Greece in 1919. He and Konomos (1986: 123–6) both maintain that the old exploitative system of land tenancy was legally circumscribed by the Zakynthian politician Nikolaos Kolyvas in the 1920s.

11 In mainland Greece a series of agrarian reforms in the inter-war period succeeded in breaking up the large estates and instituting 'the small peasant-holding as the dominant form of land ownership' (Mouzelis 1976: 87; 1978: 91, 79). Vassilikos was less attuned to those developments.

12 Xenopoulos 1936, 1945, 1959a, 1959b, 1984; Roma 1967, 1971, 1973, 1975, 1980; Konomos 1981, 1983, 1985, 1986.

13 To give some credit to this lady, I admit that, for any social scientist, an interview with her would have been astonishing and priceless. A generation of Zakynthians of her age are vanishing, along with valuable unrecorded life-histories and memories, capable of illuminating varying aspects of Zakynthian life in the early part of this century.

14 The Sea Turtle Society (STPS), the Mediterranean Association to Save the Sea Turtles (MEDASET), the Ecological Movement of Zakynthos (ZOK), WWF-GR and Greenpeace-GR (see also Chapter One).

15 See, for example, the competition between STPS and MEDASET described by Botetzagias (2001).

16 Loggerhead Sea Turtle, *Caretta caretta.*

3
CONSERVATION AND THE VALUE OF LAND

This chapter examines the Vassilikiots' intense and intimate relationship with the land, 'their land' (*ti gi tous*). Divergent sets of cultural meaning – such as, the personal significance of land ownership in a community where land has been a scarce resource, the importance of land for the development of tourism, the association of the family name with particular plots of land and the value attributed to the notion of toil or 'sweat' embedded in the land – figure prominently in the Vassilikiot discourse. In the last twenty years, however, environmental conservation has challenged the immediacy and intimacy of the human-land relationship in Vassilikos by imposing restrictions on the freedom of some local landowners to develop tourism on their land. In their attempt to put into words their opposition to environmental conservation, Vassilikiots draw examples and metaphors from their own engagement with the land, skilfully recombining different sets of cultural meaning about the land to form new arguments in support of their case.

An extensive corpus of anthropological literature testifies to the abundance of meaning and cultural signification attributed to the land and land ownership by farmers in the Mediterranean region. Ownership of land is perceived to imply security, independence from affines or employers (Davis 1973: 73, 94–5, 161; Lison-Tolosana 1966: 67–72, 159–60; Loizos 1975: 50, 55, 61), the substantiation of a local identity (du Boulay 1974: 21, 32, 161; Pina-Cabral 1986: 126; Herzfeld 1985a: 57–8; Gefou-Madianou 1992b: 114). Land is the basic prerequisite for realising self-sufficiency, a necessary prerequisite for the successful establishment of the rural household (Davis 1973: 94; du Boulay 1974: 36–7; Pina-Cabral 1986: 63–5; Loizos 1975: 48). It is also the spatial terrain vital for uniting the economic activities of family members and maintaining the coherence of the household (Davis 1973: 45–6, 161; Lison-Tolosana 1966: 39–53, 155; du Boulay 1974: 32–40). Status, respect, political power, and social stratification are all related to land ownership (Lison-Tolosana 1966: 49, 62–72; du Boulay 1974: 176, 248, 251; Loizos 1975: 48, 111, 114, 310; Pina-Cabral 1986: 25, 29, 35, 152–4; Herzfeld 1985a: 43–7, 311), while marriage strategies pay serious attention to it (Lison-Tolosana 1966: 16, 155, 159; Davis 1973: 36, 40, 73, 159; Loizos 1975: 48, 100, 312–6; Pina-Cabral 1986: 53, 57, 63–5; Sant Cassia 1982: 649–53). Finally, cross-generational inheritance, labour and cultivation of identical plots of land provide symbolic connections between landowners and their ancestors, generating perceptions of continuity between past and present village life (du Boulay 1974: 21, 39, 139–40; Loizos 1975: 100; Pina-Cabral 1986: 126; Seremetakis 1991: 28, 43; Gefou-Madianou 1992b: 114–5; Just 2000: 190–211).

Those multiple manifestations of symbolic and material capital ascribed to the land, along with the traditional farming lifestyle upon which they are founded, are often challenged by recently introduced urban definitions of status, wealth and personal achievement, such as education, urban employment and new patterns of consumption (du Boulay 1974: 175–6, 251–2; Sant Cassia 1982: 653–61; Stewart 1991: 126–7; Argyrou 1996). In those cases, however, where land provides the basis for viable economic exploitation or development, as in the community examined in this book, traditional notions relating to the value of the land continue to provide powerful statements about the identity of the landowners. In Vassilikos, new sets of land valorisation, like those related to tourism development, coexist in parallel with the older cultural ideals, which underline the symbolic and material significance of the land for the farming economy. In fact, both categories of signification are often employed by local actors, critically, selectively, and in some instances jointly, in order to safeguard particular collective or individual goals in varying contexts of social life.

As I will describe in detail in this chapter, the multiple meanings of land ownership in Vassilikos are the result of painful experiences of landlessness in the past and an immediate engagement with the productive resources of the land (tourism and farming) in the present. In the context of the environmental dispute, those meanings are creatively reunited to form a constantly transforming discourse, which reflects the unwillingness of the local landowners to be parted from their land. My ethnographic account begins with an overview of the social and material circumstances faced by the inhabitants of Vassilikos at the time when they were landless tenants (*semproi*) on the estates of wealthy landlords (*afentadhes*). In the subsequent section, several Vassilikiots describe in their own words their long and painstaking efforts to gain access to and ownership of land. Following this, examples of land valorisation in Vassilikos are discussed in detail, in an elaborate account of the cultural depth and richness pertaining to the indigenous relationship with land. The last section examines the imposition of land conservation measures in designated areas of the local environment and the rhetorical attempts of the landed actors to bring together composite versions of land valorisation in their attempt to retain control over their property.

The time of landlessness

Throughout the Venetian era (1485–1797), and even later, when Zakynthos was a British protectorate (1815–1864), a small number of noble families, known in Zakynthos as the *nobiloi*, controlled the greater part of fertile land on the island. Landed property usually remained undivided, since the old inheritance custom of exclusively favouring the firstborn son was prevalent among the aristocracy. The *popolaroi*, the numerous merchants and artisans residing in the island's capital, challenged the power of the aristocracy on several occasions, but their ideas were hardly felt in the countryside. The inhabitants of the countryside were mostly landless tenant labourers who remained, as most Zakynthians agree, persistently 'faithful to their masters' (*afentopistoi*). They were all – even the few who owned some small pieces of land – referred to as *semproi* by both the *nobiloi* and the *popolaroi*. This term could be literally translated as 'peasant tenants', but in Zakynthos it was associated with the feudal past and connotations of serfdom. For town dwellers, aristocrats or bourgeois alike, the *semproi* of the countryside were treated as 'illiterate and unmannered peasants' (*agrammatoi kai axestoi khoriates*) (c.f. Hannell 1989: 110).

After the unification of the island with the Greek state in 1864, the aristocrats lost power to the wealthiest of the urban middle class. However, the social emancipation of the latter had little effect on the lifestyle of the inhabitants of the countryside. Most of them remained landless or had very small land-holdings to sustain their households. They had to work the land of some powerful others as semproi; but now the majority of these powerful others were landlords of bourgeois descent. In all cases, the *semproi* were entitled to a small percentage of the produce, which was determined by the elaborate system of *sempremata*, according to which standardised forms of economic arrangements between landlords (*afentadhes*) and labourers (*kopiastes*) were applied to different kinds of productive resources (cultivation or animal husbandry). The individual arrangement between those two parties, the landowner and the tenant farmer, was locally referred to in Zakynthos as a *sempria*.

In the years preceding and following the Second World War and until the early 1960s, two thirds of the cultivated land in Vassilikos was part of a single, old-established estate. The legitimate heirs of this estate were two brothers, members of an old, noble Zakynthian family. The older brother was named and referred to by the villagers by his title, 'the count' (*o kontes*). He was the master (*o afentis*) of the land and the tenant farmers (*oi semproi*) living on his estate were exclusively dependent upon him. In the 1960s, his property was inherited by his nephew, an educated man who disapproved of the noticeable remnants of feudalism in the village. He sold off plots of land at relatively low prices to local farmers, who had been working on the estate of his father and his uncle for many consecutive years. Despite this disposal he still owns most of the land in the area, since by being the only heir of the estate, he inherited a large amount of land. Most of the villagers – but especially the older ones – still treat him with a kind of respect reminiscent of social stratification in the past.

During the same period, the rest of the cultivated, fertile land in the plains of Vassilikos was owned by landlords of upper middle-class origin, wealthy people living in the capital of the island. In the latter part of the twentieth century, most of these urban landlords lost or sold their land in Vassilikos. Their landed property was divided into smaller plots inherited by numerous descendants, a testimony to the power of the Greek inheritance law, which has always encouraged land fragmentation to discourage the proliferation of large estates (Herzfeld 1985b: 167–83). In the past, these smaller landlords, despite their bourgeois origin, employed the pre-existing system of rights and regulations (*sempremata*) in respect of the cultivation of their land. Their land was cultivated by tenant farmers (*semproi*) according to arrangements identical to those used by the landed aristocracy in the past. Like the aristocrat landlords, this second category of landowners were approached by their tenants with a combination of respect and fear. The latter referred to their bourgeois landlords by the term 'master' (*o afentis*), while their attitudes and manner of interaction with them was indicative of deference.

The life-histories of the older Vassilikiots contain vivid memories of landlessness and dependence on wealthy landowners for access to land suitable for cultivation or animal pasture. This period in their lives is described as the time when they were landless *semproi*, working and living on the land of the landlords (*afentadhes*). This is how a seventy-year-old man talks about that time:

> Most of the time, the landlord used to place you on some piece of land (*khtima*), according to the size of the family you had; for example, how much land you could cultivate. Some times, the landlord would replace his *semproi*; for some reason he might not want them to stay. In this case he could give them three months notice to find another place. Sometimes, though, one family could have stayed in the same place for many years...

A younger, fifty-year-old man, further explains:

> Many families used to stay on the same plot of land for years. Often, sons were cultivating the land that was previously cultivated by their fathers. But this was not *their* land. It was the landlord's land (*tou afenti*). He used to ask them to sign a contract every four years, declaring that they had just arrived on his land. In this way they couldn't claim ownership of the land.

The periodic movement of particular families of tenants or *semproi* from one plot of land to another – often within the confines of the same estate – was intended to safeguard the property rights of the landlords against potential land claims raised by the tenant labourers. According to state legislation if someone is 'using', that is living on or cultivating, someone else's land for a period of more than twenty years he may claim ownership of this particular piece of land (cf. Handman 1987: 51). This law is referred to in the Greek legislation as *hrisiktisia* and accounts for those cases where the Zakynthian landlords insisted that the *semproi* should move to another piece of land or sign a contract stating that they were new to the land. Du Boulay (1974: 269–70), in her classic ethnography of a Greek mountain community, refers to this law by its formal name (the Law of Possession: *Nomos Katohis*) and states that her informants, like my Vassilikiots, considered it to be a source of conflict and enmity among fellow villagers.

Finally, it must be noted that mountainous areas with land less suitable for intensive cultivation, unlike fertile land on the plains, never attracted the interest of landlords. In Xirokastelo, for example, an area which is administratively part of the modern community of Vassilikos and geographically adjacent to it, the mountainous character of the terrain discouraged the wealthiest Zakynthian lords from retaining or incorporating parts of this land into their estates. Some of this mountainous land was – and still is – monastic property owned by the Monastery of Saint Dionysios. Monastic land, like land on private estates, was often cultivated by tenant farmers (*semproi*) according to the system of *sempremata* practised elsewhere on the island. But those Vassilikiots who have worked on monastic land unanimously admit that monks were much more lenient than the landlords on the plains of Vassilikos. 'The Saint (*o Agios*)', they add humorously, 'has always been a good master!'

Narratives of land acquisition

In the mountainous area of Xirokastelo, the most inaccessible part of Vassilikos, I recorded the oldest cases of ordinary farmers owning land in Vassilikos. Taking advantage of the inattention of the state or the aristocrats, some poor families managed to clear neighbouring bush and enlarge their smallholdings. This strategy of clearing scrubland, a practice that thrives in the absence of systematically recorded land titles and other legal documents, is well documented among Greek speaking country dwellers (cf. Handman 1987: 50–1; Herzfeld 1985b: 169, 181; Sant Cassia 1982: 649). In Vassilikos, memories of 'clearing bush' refer to a shadowy and unspecified era in the past, the time when state legislation had little restraining power on the everyday practices of landed or landless actors.[1] In fact, the subject of past clearances of deserted, uncultivated lands always fascinates men and women in Vassilikos:

Some families at Xirokastelo always had some land of their own. Their land used to be scrub (*lagadhia*) and they cleared it (*ta xekhersosan*). No one knows exactly how they got this land.

Nowadays, it is forbidden to cut scrub. In the past people used to find empty stretches of scrubland and cleared them; as did the people of Xirokastelo, for example.

The land obtained by this method of clearance was not very suitable for cultivation. Despite this fact, the families living in the area had an opportunity to escape landlessness or total dependency upon landlords for access to land. If this kind of mountainous land were not enough to provide one with a livelihood, cultivating monastic land according to the established system of *sempremata* was an additional available possibility.

The first families of semproi to obtain ownership of land in the plains of Vassilikos were ex-tenants rewarded by their landlords. After working the same plot of land for years, some landless labourers were granted some of this land 'as a *quarto*', The Italian term *quarto*: a quarter, indicative of the Venetian origin of the system of rules relating to land cultivation, refers to the actual size of land given by the landlord to the labourer. It was a quarter of the land cultivated by the tenant; not a quarter of the actual property of the landlord. The lucky beneficiaries of this donation were faithful *semproi*, whose families had 'served the landlord well' during one or two generations. Vassilikiots remember several examples of local families acquiring land in this way:

> Some families at Dimareika and Potamia [placenames] had land of their own. They got their own land as a *quarto*, for serving the same landlord (*ton idhio afenti*) for several years. Still, because their land was not enough, they got additional *sempria*-arrangements with the same landlord.

It must be noted, however, that the mechanism of *quarto* operated within a context of long-term, carefully maintained patron dependency. Some families of faithful *semproi* were rewarded for their services with some land, which, in most cases, was not enough to provide them with a livelihood. They had to resort to their landlord to obtain the right to cultivate additional plots of land according to the established system of *sempremata*. Consequently, land donations of the *quarto* type, strengthened, rather than undermined, the patron-client relationship between proprietors and tenant farmers. Complete landlessness was avoided, while the villagers were further enchained by obligation to their patrons.

In the years following the Second World War, there was increasing pressure on landlords holding big estates to sell or distribute plots of land to their landless farmers. This situation had an effect on Vassilikos, which due to its geographical isolation, had been less attuned to the social changes occurring in other parts of the island since the beginning of the century. Some Vassilikiots have reported incidents of landlords being murdered in neighbouring Zakynthian villages in the late 1940s. The civil war which took place in mainland Greece between left and right-wing fractions contributed to the creation of an atmosphere of general confusion, within which social tensions at the local level were resolved by murder (cf. du Boulay 1974: 238–42; Handman 1987: 59–60, 62–3; Hart 1992: 78–80). Landlords were killed by exploited landless labourers and vice versa. The older people in Vassilikos refer to the murder of a 'leftist' landowner, the only wealthy landowner who appears to have been affiliated to the political left:

> This man made a lot of money in America as a migrant worker. He came to Vassilikos and bought an estate with a beautiful country mansion from one of the old landlords. He had learned about communism in America and he was 'educating' the peasants. The other powerful people didn't like this. He was shot in an ambush on his way to the village at a turn in the main village road.

In the same period, one of the most powerful landlords of Vassilikos was unsuccessfully attacked and shot, this time by people presumed to be leftists. 'A man approached him while he was sitting in the barber's chair', some Vassilikiots remember; 'the man placed a pistol at his temple and fired once.' Miraculously, the landlord survived. As the Vassilikiots recount with amusement, the old pistol's barrel was touching his head and the bullet had not enough power to penetrate the landlord's skull: 'The bullet was jammed in the bone!'

Vassilikiots argue that those incidents made the landlords insecure enough to start selling their land. This argument rests on the popular assumption that, the more landless people there were in the village, the greater the likelihood for dissatisfaction culminating in social unrest. During the late 1940s and early 1950s, the growing local demand for land ownership became overwhelming. Most landless labourers became increasingly aware that the old exploitative regime pertaining to land tenure was not to be tolerated in the mid-twentieth century. Some farmers, but especially those of leftish persuasion, consistently criticised their fellow villagers for being 'faithful-to-the-master' (*afentopistoi*). These are the words by which a fifty-year-old Vassilikiot refers to this period:

> The landlord was compelled (*anagastike*) to sell land to the people at reduced prices, for example, twenty thousand drachmas instead of a hundred thousand. He sold the farmland (*khtimata*), which I bought, and the land that all the others hold in this area. If he had done otherwise, they would have killed him (*tha ton trogane*). His uncle [the previous landowner] was shot at three times (*treis smpares eikhe faei o theios tou*).
>
> Nowadays, many people think that they benefited from him (*tous ofelise*) and they pay respect to him. But still, he has so much land! Vassilikiots were among the most 'faithful-to-the-master' (*afentopistoi*) people on the island.

Another, younger man, locally known by the nickname 'Ringo' – a statement about his overt masculine character and behaviour – was fearless enough to admit:

35

This land that I have – it is not even one *strema* [a quarter-acre] – it was given to me for free (*mou tin harisane*). I told them, either you will give me a little land or I will become a thief. This is how I got this little piece of land.

Unlike Ringo, most of the people in Vassilikos had to work hard in order to secure a minimum amount of cash to buy small pieces of land. Some of them had to migrate to mainland Greece and work as manual labourers for years, intimately associating their plans for a successful return with the hope of land acquisition. The following narrative by a sixty-five-year-old man in Vassilikos illustrates this point:

My family originates from Volimes [a mountain village in Zakynthos]. They were forced to leave Volimes and went to live in a marshy, poor area in Kalamaki in the Zakynthian plains. This area, now, is the centre of tourism. Some tough bullying shepherds (*tsampoukadhes voskoi*), with guns, trampled down their crops and forced them to leave and become *semproi* in Vassilikos. This is where I was born, at Kotronia. My father and his brother didn't succeed in buying land and they got separated. My father went to Xirokastello. He worked as a *sempros* on the Saint's land.[2]

But since making a living was hard at the time, my father went to mainland Greece to work as a gardener. Zakynthians, you know, used to be renowned for their skill in gardening. My father made some money in this way.

We bought this land from the landlord in 1953 with 60,000 drachmas paid in English pounds (*se lires*). It was important that this money was in pounds.[3] The landlords were in need of cash at the time. They were used to gamble (*tzogaran*) at the Casino, hoping that they might win; but they were always losing! Another landlord [he refers to a well-known rich Zakynthian] found them in difficulty and he bought land from them (*tous vrike se dhyskolia kai tous pire gi*). Then my parents heard that land was for sale in Vassilikos. They rushed back to Zakynthos to learn more about it. I was crazy with happiness when I saw this piece of land (*trelathika ap' tin khara mou*). We started planting olive trees.

During the 1950s and 1960s some Vassilikiots managed to secure plots of land, while others persistently failed to do so. Some had a few opportunities, but, as they explained, their hesitation to obtain land stemmed from their fear of or respect for their landlords. Others failed to acquire land due to a variety of reasons relating to their passion for gambling, drinking, or other personal indulgences. The following account by a sixty-five-year-old man colourfully touches upon those issues:

At this time Tsagkaris [a nickname] bought some land in Vassilikos. He already had some land as a *quarto* at Doretes. He had a lot of goats and animals of this kind. Everybody in his family worked hard and managed to buy some more land. They were among the first Vassilikiots to have land of their own.

Most of the people were offered some land by their masters but they didn't want to accept it. They were afraid. They used to say: 'Is it right, my master, for me to have land? How I will be able to look you in the eye?'[4]

Those people were very faithful to their masters. They were denying themselves, not stealing, not even one *ogia* [a weight unit] from their master's olive oil. For example, if the olive oil was 31 *ogies*, they used to say 31, not 30. They would say: 'Shall *I* steal from *my* master?'

To others, like the father of Nionios who was the overseer (*epistatis*), the landlord would offer a little piece of land. He was always refusing to accept it. He said, 'I live on your property, master, you feed me and you keep me alive, why should I need land of my own?'[5]

Probably, those people were afraid because of the stories about the '*narkova*' [a deep pit]. It was said that sometime in the past the *semproi* were told by their master to come to the town so as to be given some land. They went to the town for the contract, but they were thrown into a deep pit (*khantaki*), which was covered like a trap. They were told by the master: 'Come here!' and they fell into the pit. Then, the master would say that the dead *sempros* had gone to America as a migrant, or the master would ask his relatives: 'Where is he? I was waiting for him, to give him some land.'

And here is another shorter example, narrated by a local woman in her late forties:

Veniamin [a nickname], Mimis's father, lost the money which he and his wife had been saving for years in order to buy some land. They sold cattle and animals so as to collect the required amount of money. Then, he went to the town to sign the contract. But he was tempted to gamble with the money and he lost it all (*ta epaixe sto tzogo*).

While most of the older Vassilikiots consistently recount their endeavours during the years of landlessness and proudly point towards their recently acquired land, some younger individuals do not appear particularly impressed by their parent's narratives of toil and success – stories they have heard repeatedly in their childhood. 'My children cannot realise how lucky they are to have inherited land from us' an elderly woman frequently complains. Having had several intimate conversations with her 'unappreciative' children on that issue, I can testify that the latter are very conscious of the value and importance of their family land. What the younger Vassilikiots do not always acknowledge is the strategy of exhausting manual labour employed by the older generations during their lifelong struggle to secure their own land.[6] Referring to a middle-aged man, who works extremely hard in his perpetual effort to buy further pieces of land, a young Vassilikiot commented: 'He makes his life less comfortable (*mizerevei tin zoi tou*), the clothes he wears for example, so as to buy more land from the landlord every year.'

The person criticised by the young man in the previous quotation offered me his own, different perspective. He maintains that he feels a great deal of injustice about the inequality in land distribution. On several occasions he pointed out to me land neighbouring his own which was the property of the landlord (*tou afenti*). Then, he compared his painstaking efforts to 'make use' of every small piece of his own land, with the relative under-utilisation of the landlord's large amount of land, which is cultivated less intensively, and remarked:

> My father, although he was a communist, did not achieve any prosperity (*dhen prokopse*)... He didn't buy any land... He was talking 'ideologies' all the time. But I am not satisfied with my own progress either. When I was younger, I could have done more... I could have got more land... but my wife was always stopping me... she was asking me not to wear myself out... When I was younger I could wring water out of a stone. I worked so hard and I deserved more.

Finally, the same man explained to me that having been born landless, his success in escaping from 'the fate of the tenant farmer' (*ti moira tou semprou*), was an event of great personal significance to him.

All the examples presented in this section illustrate the importance attributed to land ownership by the people of Vassilikos. For some of them, acquiring land of their own was the realisation of a lifelong goal and the result of persistent efforts. Within a period of forty years, they went from of a situation of complete landlessness, to a comparatively more comfortable economic position. Nowadays, almost every Vassilikiot possesses some land suitable either for cultivation and animal husbandry or for small-scale tourist enterprises. Several local farmers still cultivate land owned by landlords or the landlords' descendants, according to patterns of *sempremata* which have been modified so as to allow greater profit for the labourer (*kopiasti*). Most of the local people wish to expand their landholdings so as to allow for more productive economic activities, related either to animal husbandry or tourism. For them the struggle to acquire land is a process which has not yet been fully completed.

The value of land in Vassilikos

The people of Vassilikos talk a lot about their land; they talk about it poetically, emphatically, metaphorically. The rhetorical tenor of their statements is highly dependent upon the particular context in which the discussion takes place. In one conversation, for example, the Vassilikiots would praise the fertility of a specific plot of land and its suitability for agriculture and animal husbandry. In another, they would assess the potential of the same plot of land for tourism development, by examining its proximity to the seashore, the main road, or other tourist services. In a third instance, a discussion about the land, the land owned by a Vassilikiot family, might serve as the starting point for unravelling the personal significance of this land as the foundation of the family's independence and future prosperity. But while different occupational identities – such as that of the farmer or the tourist entrepreneur – are often negotiated by the same individuals in diverse situations or social settings, the value attributed to the land remains equally significant in all cases. In fact, most Vassilikiots rarely differentiate between the different sets of significance they impute to their land. They simply stress particular meanings in specific conversations.

Above all, Vassilikiots understand land ownership as an achievement. As I have already described in the previous section, the great majority of the landless labourers in Vassilikos acquired 'their own land' (*tin dhiki tous gi*) with painstaking efforts, over the last thirty or forty years. Before this, fertile land was a precious, scarce resource, available only to a few privileged families. Prolonged landlessness and dependence on landlords for access to land infused the local meaning of land ownership with a powerful emotional and symbolic content. For most people in the village, land ownership used to be a lifelong aspiration, the major objective of their hard-working life. Having now realised this ambition, they can sit back, admire their property and reflect on the mark their labour has left on what is now 'their land'.

Vassilikiots literally read their life-histories by staring at individual plots of land. 'Their own land' (*i dhiki tous gi*) is the reward for being patient and diligent during the times of landlessness and poverty. It is the tangible evidence that all the 'sweat and toil' (*o mokhthos*) they have spent, has not been wasted or 'lost' (*dhen khathike*). It is proof that they have escaped the fate of the landless tenant (*tin moira tou ftokhou semprou*) and have a place, their place in the sun (*ston ilio moira*). This is why land is for the Vassilikiot farmers a sign of self-determination, the solid foundation of their current endeavours for 'a better life'. It embodies both their past efforts to escape from poverty and their hopes for the future in the present.

Land ownership is also important for the establishment of a strong sense of local identity. To have access to landed property in Vassilikos is one step on the way to becoming a Vassilikiot. To reside or work on that property over a long period of one's life is a second. To be able to trace cross-generational kinship links in the locality is a third and more significant step. Thus, the perpetual presence of the same inhabitants, preferably people with the same surname, on the same plots of land gives birth to a sense of symbolic continuity (cf. du Boulay 1974: 21, 39, 140; Seremetakis 1991: 28, 43; Gefou-Madianou 1992b: 114–5). The strong association of land with 'the name of one's family', fashions the local environment into plots of land where the presence of particular families is synonymous with the land itself. In this way, the legitimacy of land ownership is reinforced, and any possible lacunae in the formal documentation of landed property are easily resolved.

The association of family names with particular plots of land has been facilitated by Vassilikiots' preference for a patrilineal pattern of land inheritance. Until recently, most local girls traditionally received money or other forms of movable property as dowry, while land was normatively inherited by male descendants of a family (cf. Gefou-Madianou 1992b: 114; Just 2000: 126, 191, 211). A fifty-year-old Vassilikiot elaborates on this issue:

> Girls were never expected to inherit the land of their father (*tin patrogoniki gi*). If land was to be given to them as dowry – this could have happened in the case where the groom had no land – land was bought for them. But family land had to remain in the name of the family.

I was initially surprised to encounter such a clear-cut articulation of the normative ideals on inheritance in a community where land acquisition was a relatively recent phenomenon. I soon realised, however, that in Zakynthos a strong patrilocal ideology has been dictating the rules of land inheritance since the period of Venetian rule. Native novelists and historians have repeatedly referred to some Zakynthian aristocratic families who allowed only one of their male descendants to marry and procreate in order to prevent the division of the family's landed property (cf. Roma 1971, 1973). The oldest among Vassilikiots described similar customs practised by families of farmers in the mountainous villages on the west side of the island.[7] 'The people in the mountain villages', they explain, 'used to marry off (*na pantrevoun*) only their younger brother because they hadn't enough landed property'. The general idea behind this custom was that the older, unmarried brother would remain the head of the household, while the younger, who was supposed to tolerate his brother's authority, would eventually enjoy the privilege of 'seeing' (*na dhei*) his son inheriting the household property. Nowadays, Vassilikiots, like all other Zakynthians, consider this kind of inheritance stipulation to be obsolete.

On the other hand, several Vassilikiot men clearly express their distaste for matrilocal residence. Being *sogabros* [an in-marrying son-in-law] is considered by most of them very shameful (cf. du Boulay 1974: 126). 'Even a poor man would prefer to live away from his parents-in-law'[8] they argue. But women express a similar, yet diametrically opposite position on this issue. On several cases they described to me the psychological 'pressure' (*piesi*) they experienced, when they realised that they had to abandon their paternal household in order to 'live with' – and 'put up with' – the oddities (*paraxenies*) of their parents-in-law (cf. Hart 1992: 173–4).

During my fieldwork days in Vassilikos I recorded several cases of girls inheriting landed property. Some men in the village felt threatened by those instances. Once, I heard one young man saying to another: 'you burn our fingers (*mas ekapses*) by giving land to your sister'. A middle-aged woman explained to me that the young man who complained had a sister as well. He was pressed to accept the possibility of his sister inheriting some family land. This example demonstrates that patterns of neolocality infiltrate into the local society, dictating new forms of land inheritance. Numerous recent exceptions to the normative patrilocal ideals indicate that Vassilikos is undergoing a change in respect to issues of land inheritance and postmarital residence. According to the model offered by Loizos and Papataxiarchis (1991a: 8–10), Vassilikos can be accurately described as a community in transition from patrilocal rules of residence, with a strong emphasis on agnatic descent, to neolocality and bilateral rules of inheritance.

Most Vassilikiot men and women – and especially those of the older generations – maintain that land ownership is one of the foremost qualifications of a good marriage partner. In accordance with the traditional views on patrilocality, land ownership enhances the marriageability of young men, since it provides spatial and economic independence from one's affines. But, even under the new 'somewhat bilateral' patterns of postmarital residence, landed property is considered to be a primary, fundamental resource, upon which the married couple can base a new family. The more general ideal – widespread in rural Greece – that a newly married couple comprises an independent economic and social entity (Loizos & Papataxiarchis 1991a: 6), reinforces the local appreciation of landholding as a vital prerequisite for a successful marriage. It must be noted, however, that in Vassilikos the independence of a newly founded household is more often measured uni-directionally in respect to the groom's affinal group. Several married Vassilikiot women reside patrilocally after their marriage and make a living out of the groom's family land, which both marriage partners consider with confidence as 'their own land'.

Being fully Vassilikiot, with well-established kinship roots in the area and ownership of some land, is a criterion that renders access to a further set of resources, those related to tourism. This does not mean that outsiders are completely excluded from tourist enterprises (cf. Galani-Moutafi 1993: 250–1; Kenna 1993: 87–8). Various non-local people find their way into the business of tourism, owing to their close relationship (kinship or friendship) with the locals or their own personal skills (knowledge of foreign languages, music, bars, or other forms of entertainment). But those frequent examples of non-local people involved in tourism in one way or another are treated in Vassilikos as 'exceptions' rather than the rule. The clearly articulated position of Vassilikiots in respect to this issue is that non-local people wishing to make profits out of tourism are not welcomed (cf. Stott 1985: 204).

On the other hand, Vassilikiots' entry into the economy of tourism carries an aura of legitimacy. This is because tourism is considered to make permanent residence in the village viable and justifiable. It is regarded as a benefit, a reward compensating for comforts or economic resources that the village life lacks. It is also perceived – especially by the poorer Vassilikiots – as the best locally available alternative to poverty. Land ownership, like local identity, entitles one to access this benefit, since it is locally claimed that all people, who share a bond with the land of Vassilikos, deserve (*axizoun*) to benefit from it.

Unlike in the recent past, when traditional farming was the primary economic strategy in Vassilikos, tourism nowadays provides the greater source of income for most local inhabitants. This fact does not diminish the earlier material significance of land for the local people. On the contrary, any villager aspiring to enter the tourist industry by means of any form of legitimate or reliable enterprise needs access to landed property. Consequently, the value of land has been increasing along with the development of tourism. Plots of land closer to the beach or to the village main road gain additional value, since they provide ideal settings for various tourist enterprises. The older Vassilikiots frequently reflect on how land unsuitable for cultivation has now obtained a new kind of value. People owning property close to the sea, for example, had once been considered as unjustly disinherited, but, nowadays, their 'infertile' (*agoni*) land on the seafront is the most fertile terrain for developing tourism (cf. Herzfeld 1991a: 41, 73, 154; Boissevain 1996b: 10; Zarkia 1996: 150). As one Vassilikiot man puts it:

> In the past, we used to say 'they gave us just a bit of sand' (*ena kommati ammo*), suggesting that land 'down by the sea' (*kato stin thalassa*) was given to an unlucky person who was disliked by his relatives. This kind of land has sandy earth, where nothing can grow.
>
> But now the terms have been turned around. Now, some people see what has happened and pull their hair out!

Regardless of the particular location, however, almost all land in Vassilikos is potentially suitable for the development of tourism. Even the most isolated areas lie within reasonable driving distance of the main beaches, which are the focus of tourist activities. Consequently, it is not surprising that the owners of this kind of relatively inaccessible property retain some realistic and several unrealistic aspirations to develop their land in one tourism-related way or another. Landholding in Vassilikos embraces a spontaneous claim to participate in the business and benefits of tourism and all local land owners feel justified in investing emotionally in a future of tourist development.

Finally, apart from being a prerequisite for exploiting tourism, land is perceived by Vassilikiots as the safe foundation from which one can embark on insecure ventures such as this. 'Tourism doesn't always do well' (*dhen vgenei*), the local protagonists maintain. This is why one's own land is locally described as one's bedrock for dealing with unpredictable risks or lack of success (cf. Davis 1973: 161). Alternative sources of income based on land, such as cultivation and animal husbandry, are the safety valve for fluctuations in the tourism economy. Most Vassilikiots still practise traditional farming activities. Some 'keep' animals and make considerable profit out of cheese making, others retain large olive groves, construct greenhouses for growing vegetables or cultivate vegetables (especially tomatoes and melons) in summer season gardens in the open air. The vast majority of Vassilikiots maintain the identity of the farmer and, like all farmers, proudly depend on their land. As they put it, 'if tourism doesn't do well, the land will be here for us'.

Conservation and resistance

During the last twenty years a series of presidential decrees and state laws dictated the creation of a Marine Conservation National Park in Zakynthos. The Marine Park in Zakynthos includes parts of the coastal environment, and in particular the south coast of the Vassilikos peninsula, the most underdeveloped part of the community. This is where several local families, related by kinship ties, own land that is directly affected by the conservation restrictions. The land to be conserved is relatively inaccessible and, unlike other parts of the Vassilikos peninsula, little tourist development has taken place. In addition, the terrain is steep and does not allow for intensive cultivation. In the last quarter of the twentieth century, however, local landowners realised that improving the existing dirt road could lead to possible development of the area, given that the local beaches – the same beaches where the turtles lay their eggs – are of substantial natural beauty. Thus, small-scale tourist enterprises in the form of fish-*tavernas*, umbrella and canoe renting, started to establish themselves from the 1980s onwards. The local landowners lack the capital to invest in grand projects, but having tasted the profits of tourist-related enterprises, they visualise the future development of their land as being inextricably linked to tourism.

Evidently, the Marine Park constitutes a serious obstacle to the fulfilment of the local landowners' visions for economic development. From the point of view of the Vassilikiots, ecological conservation was instituted unexpectedly in 1982 and 1984,[9] when a couple of presidential decrees prohibited building on the land adjacent to the turtle-breeding sites. Three years later, a new law further reinforced the previous restrictions. Small-scale tourist enterprises on the turtle-beaches, like the ones that were already in operation, were circumscribed and any human presence on the beach during summer nights was strictly forbidden.

Despite their apparent severity, the conservation measures of the Marine Park were never properly implemented in Vassilikos. After waiting in vain to be compensated for their expropriated property, the affected landowners collectively declared their opposition to the national park and harassed – by constant threats and, in some cases, physical violence – the various groups of conservationists attempting to gain a foothold on their land. On several occasions the police and other civil officials attempted to stop the erection of illegal buildings constructed on the conservation area. They always returned to their headquarters spectacularly unsuccessful. The local spirit of resistance dramatically displayed in stances of 'performative excellence' – to quote Herzfeld (1985: 16) – successfully undermines the reluctant efforts of the local authorities to impose the legal conventions. In the meantime, Vassilikiots appear determined to exercise their will, which is locally perceived as a 'right' (*to dhikaioma*) to 'do whatever they want to do with their own land'. Narratives such as the following were often heard in the village, during the time I was conducting my fieldwork:

They tried to pull down the new illegal constructions (*afthaireta*) in Dafni today. But one of the owners (his name is explicitly stated) was waiting for them. He went down the road with a gun and he stood in front of the bulldozer and the Public Prosecutor. He said: 'Get down, if you dare (*opoios einai antras as katevei kato*). You will not pull down my house on my land, which I own with legal papers. Come on, give me back the taxes for the purchase. Why didn't you stop me, when I was paying the taxes?'

In the early 1990s, WWF International succeeded in buying some land surrounding one of the three turtle-beaches in Vassilikos at a substantial price. The owners of this land declared that they did not wish to sell their land, but, tired of the long wait for appropriate compensation, they eventually had to accept the offer and sell their land at a decent price. 'What's the purpose of keeping land', they said, 'if we are not allowed to have adequate control over it?' Other local people owning land in the conservation area disapproved of them selling land to the conservationists. One of them explained to me in a deep persuasive voice: 'I will never sell my land. Look at *this man*. He sold his land to WWF and now comes to my place to fish and moor his fishing boat!'

The same man, who declares that he will never sell his land, will probably sell it – his fellow villagers maintain – if he is offered the right amount of compensation for it. Most of the landowners in this position reside on and own plots of land in other, less marginal areas of Vassilikos, which are not included in the national park. It is unlikely, therefore, that they will ever remain landless or homeless. Fair compensation will free them from anxiety, uncertainty and the endless struggle with the conservationists. On the other hand, in the absence of any form of compensation, and in the face of continuous intervention from outsiders, the landowners affected by conservation measures in Vassilikos have every reason to oppose the conservation measures in their persistent effort to retain control over their land. The overwhelming majority of their fellow villagers are on their side, both people related to them by bonds of kinship, and other sympathisers, who feel committed to support their neighbours in their recent predicament.

To justify their resistance to environmental conservation and underpin the symbolic and economic value of the land under conservation, Vassilikiots articulate novel, composite versions of land valorisation, which rely on both their new identity as tourist entrepreneurs and the older one as farmers and country dwellers. The tourist economy provides them with arguments relating to the material loss incurred by being prohibited from fully exploiting the potential of their land for new enterprises. Comparisons with other areas of the island, where tourism has been overdeveloped, even at the expense of the turtles' biosphere, raises ethical considerations about a form of legislation or state policy which preferentially allows access to prosperity. From their previous life experiences as farmers who have worked and 'cared for' the land, Vassilikiots borrow equally powerful metaphors and symbols with argumentative aptitude. The bond of the cultivator with the land is emphasised along with the symbolic significance of inheritance and kinship ties. For the people of Vassilikos, land ownership entails the complete and undisputed right of the owner to control the land and manage all its potential economic or symbolic resources – tourism, like agriculture, being examples of resources of this sort.

In the battle over land conservation, different identities and different discourses of land valorisation unite and reinforce one another. The same land Vassilikiots work and pragmatically utilise, is portrayed in narrative in the most affective, idealistic terms. Multiple and divergent sets of value assigned to land ownership are presented as self-evident facts. The landed actors of Vassilikos explicitly point out that their relationship with the land – 'their land' – is a very serious one. The non-local participant in any relevant conversation is not even allowed to question this point.

> But this is my land, the land my father secured with sweat and effort. What do you expect me to do? Offer it as a present to the ecologists? (*Na tin khariso stous oikologous?*)
>
> These are serious matters (*sovara pragmata*), my dear (*matia mou*). We are talking about people's land (*ti gi ton anthropon*)... the land of our fathers, the land we work, the land we sweat... people's legal land...

Here, the multiplicity of meanings attached to the land adds to the weight of the land-actor relationship. The polysemic signification of the land is readily translated into evidence with argumentative power. In conversations about conservation, meanings drawn from land symbolism merge with practical considerations and aspirations for material and social progress (*prokopi*). As such, the ancestors' sweat and suffering embodies all current dreams for economic growth. Metaphors about the land marry tradition with development, while divergent sets of meanings, that might contradict each other in the context of other conversations[10], regroup and unite to counter the validity of land conservation.

I will conclude this section with an extract from a report written by a group of Vassilikiot landowners directly affected by the establishment of the Marine National Park in the vicinity. The report is entitled '*Memorandum of the owners of landed property at Gerakas, Dafni and Sekania in Vassilikos Community*'[11] and is addressed to the Prefect of Zakynthos. It neatly sketches the Vassilikiots' position regarding the conservation of their land:

> This land which we possess today belongs to us. It was bought by our grandfathers and our parents in 1955. They did not usurp this land from somebody else. Nobody gave this land to us for free. This land is the outcome of the labour and sweat of three generations, who lived and toiled all their lives, whose only dream was to acquire this land, their land...[12]

> We believe that the land which is owned by any villager, who is a Greek [citizen], belongs to him... Or do you think that his land belongs to the State, so as to be under the State's control and under the control of anybody chosen by any government in power?

Conclusion

It has already become apparent that Vassilikiots affectively celebrate their relationship with *their* land (*ti gi tous*) by stressing several aspects of their engagement with it and its productive resources. Landholding, indeed, gives access to both traditional farming and the new enterprises of tourism. It is also a primary qualification for negotiating a local, 'Vassilikiot' identity. Like elsewhere in the Mediterranean, land ownership plays a crucial role in shaping household identity, material self-sufficiency, moral independence, marriage strategies, prestige and political influence (cf. Lison-Tolosona 1966, Davis 1973, du Boulay 1974, Loizos 1975, Pina-Cabral 1986). In Vassilikos, due to past social circumstances – landlessness, dependence of the tenant farmers on powerful landlords – land ownership has acquired additional emotional and symbolic significance. The toil and sweat (*mokhthos*) of a whole generation of landless tenants (*semproi*) who have striven year after year to secure some land is embedded in this land. For the younger Vassilikiots the landed property which they have inherited is the solid foundation on which they can depend, so as to face, not merely the unpredicted oscillations of the tourist economy, but also new challenges, such as their confrontation with environmental conservation.

The Marine National Park, has indeed, threatened the local landowners, since it directly challenges their freedom to utilise the productive resources of their land – of which tourism is the most important. Standing united with their relatives, those Vassilikiots who are not directly affected by conservation, collectively resist the restrictions of the marine park by repeatedly emphasising the symbolic significance of their land. In a cultural context where self-interest and wellbeing are understood in terms of household-oriented priorities (Du Boulay 1974: 169–70; Loizos 1975: 66, 291; Hirschon 1989: 104, 141, 260), the constant evocation of divergent forms of land valorisation indirectly legitimises the relationship between landed actors and their landed property. A moral connection is made between the benefit from the land's productive resources and the right to access or control the land. The moral and the practical significance of the land are thus combined in a unique rhetorical form that constantly reflects on what constitutes justice at the local level. In local conversation land ownership communicates an impressive array of meaningful associations, legitimising the position of each interlocutor in relation to the others and to the world.

When Vassilikiots discuss their land, the simultaneous reference to multiple kinds of land valorisation brings about certainty rather than confusion. Different sets of values point in the same direction, and Vassilikiots are, indeed, certain. Their claim to control over their property is not perceived as merely valid; it is understood as rightful or just (*dhikaio*). Repetition empowers local argumentation, while the richness of cultural justification allows no space for doubt. God's trust in the farmer's guardianship of the land, the sweat of Adam or of the grandfathers, and the right to develop one's own land become parts of the same powerful rhetoric. Defying land conservation obtains a moral quality legitimised (in fact, naturalised) by the discursive unravelling of layers of cultural justification. In the context of Vassilikiot resistance to environmental conservation, the multiply signified, traditional or less traditional relationship with the land is celebrated in a unified, potent and elaborate discourse. Every landed actor in Vassilikos is in a position to testify to the intimate character of this relationship:

> I can't sell my land. I can't see the land of my father being sold to foreigners and especially to the 'ecologists'... I want to keep my land and make something nice on it.
>
> This is the land I worked. This is the land I ploughed with a wooden plow. [It is] here that I struggled (*edho palepsa*), [it is] here that I'll grow old (*edho tha geraso*).

NOTES

1   Handman (1987) has recorded that, in the period before the Second World War, state legislation concerning the clearing of scrubland was rather lenient. This is when the villagers of Pournari (Pouri) on Mount Pelion managed, with a great deal of effort, to clear considerable areas of neighbouring bush. After cultivating the cleared land for a period of twenty years, they were able to claim legal ownership over it (Handman 1987: 50–1).

2   By the personalised term 'the Saint', he refers to Saint Dionysios, the patron saint of the island. I have already mentioned that some land in Xirokastello is monastic property. Part of this land is cultivated by some Vassilikiots, who deliver a proportion of their produce to the monastery, according to the system of *sempremata*. They maintain that the officials of the Monastery have always been less exploitative than the lay landlords.

3   English pounds were perceived to be a stable form of currency at this time.

4   '*einai sosto, afenti, na 'ho ego gi? Pos tha se koito sta matia meta?*'

5   '*Afou zo sta dhika sou, afenti, me threfeis kai me zeis, ti na tin kano tin gi?*'

6   The critical attitude of the younger Vassilikiots towards 'wearing' manual labour will be explored in detail in the following chapter in the context of locally articulated tensions and contradictions between agriculture and the economy of tourism.

7   It is said in Vassilikos that the farmers of the mountain villages on the west side of Zakynthos have owned land of their own since the time of Venetian rule. As I have stated in the previous sections, the wealthy landlords of the plains had no interest in incorporating mountainous land into their estates.

8   '... *kai ftokhos na einai kapoios, protima na meinei makria ap' ta petherika tou!*'

9   In 1981, a previous presidential decree had institutionalised the protection of sea turtles in Greece as a species threatened with extinction.

10 Examples include discussions about the comparative advantages and disadvantages of tourist economy and farming lifestyles. For a more extensive analysis, see Chapter Four.

11 Gerakas, Dafni and Sekania are the disputed turtle-beaches in Vassilikos.

12 '*I gi afti pou simera ekhoume mas anikei... Einai agora apo ton pappo kai apo tous goneis mas apo to etos 1955. Dhen tin arpaxan apo kapoion. Kaneis dhen tous tin kharise, einai o kopos kai o idhrotas trion geneon pou ezisan kai mokhthisan me apokleistiko oneiro tin kataktisi aftis tis gis tous.*'

4

'BOTH TOURISM AND FARMING JOBS INVOLVE STRUGGLE'

Vassilikiots' relationship with the productive resources of their land is realised through a practical working engagement with it that is often confrontational. The work ethic pertaining to the everyday lives of men and women in Vassilikos derives from a more general combative attitude towards the environment, a constant 'struggle' (*agonas, pali*) with life. Work, and the physical fatigue that work entails, have connotations of effort or contest, a 'struggling' attitude noticed by several other ethnographers studying Greek workers in the countryside (Friedl 1962: 75; du Boulay 1974: 56, 1986: 154; Kenna 1990: 149–50; Hart 1992: 65–6; Dubisch 1995: 215a; and for Cyprus, Argyrou 1997: 163).

For most residents of Vassilikos the daily 'struggle' consists of work devoted to tourism enterprises and traditional farming jobs. In fact, the great majority of Vassilikiots are involved with both kinds of economic activities (cf. Cowan 1990: 37). They admit that tourism provides them with the most significant part of their annual income. But at the same time they insist on defining themselves as 'farmers' (*agrotes*) (cf. Just 2000: 54) and devote a considerable part of their labour throughout the year to traditional farming jobs. In this chapter I examine Vassilikiots' work or 'struggle' dedicated to tasks that involve physical labour in the open air, either in the context of cultivating the land or in the ongoing business of clearing vegetation around farms and tourist enterprises. I also evaluate the local farmers' assertion that both tourism and farming jobs involve struggle, shedding some light on the interrelationship between these two kinds of economic activity. My account is inspired by a rapidly expanding body of literature on the anthropology of tourism (Harrison 1992; Boissevain 1996a; Selwyn 1996a; Price

1996; Abram, Waldren & Macleod 1997) and contributes to it by presenting a case where the relative complementarity of tourism and farming lifestyles is challenged and simultaneously verified by detailed indigenous narratives and commentary.

In Vassilikos, the rapid development of tourism in the 1980s and 1990s did not render the pre-existing agricultural economy redundant.[1] Although most Vassilikiots make more profit from tourism than from agriculture, they do not appear set on severing their involvement with traditional farming activities (Theodossopoulos 1997b: 253–4; 1999: 613). While some local men and women still explore the mysteries of tourist enterprises, others carefully invest their earnings from tourism in building tourist apartments or buying land: the latter being locally perceived as the potential basis for both tourist development and further involvement in farming. Within the vicinity of the community, tourism, like agriculture, is developed on a small-scale family basis and success in both economies depends upon the recruitment of household members (cf. Galani-Moutafi 1993: 250; Zarkia 1996: 156; Welz 1999; Leontidou 1995: 93–4; see also, Macleod 1999: 448). As the ethnography presented in the following sections will demonstrate, profits or goods realised or produced by the labour investments of a particular household's member in either tourism or farming, are frequently used to supplement or reinforce the household's success in both fields.

Although it is hard to maintain an absolute distinction between 'tourist season' and 'out-of-season' (Abram & Waldren 1997: 3; Abram 1997: 30), the Vassilikiots annual work cycle can be roughly divided into two. The first is the period when tourists visit the island, which starts in mid-May and ends in mid-September. During this time, Vassilikiots try their utmost to take advantage of the economic opportunities provided by tourism. At the same time they strive to satisfy the minimum requirements of their farms or cultivations. The second period covers the remainder and greater part of the year, during which economic activities in the village are rather relaxed. This is when the majority of the Vassilikiots 'resume the more tranquil rhythm of their ordinary lives' (Boissevain 1996b: 9), devoting most of their attention to traditional farming activities. However, as this chapter will make explicit, even during this second, more relaxed part of the year, the local farmers do not neglect to look after their tourist enterprises or maintain them.

The next two sections present a stimulating account of contradictory indigenous views regarding the relative significance of tourism or agriculture and the labour invested by Vassilikiots in both kinds of economies. Then I will briefly discuss the local farmers' engagement with agriculture and the specifics of the agricultural work itself. 'Work in the fields' is examined, not merely as an economic exercise, but as an important part of the Vassilikiots' life, related to their identity as 'farmers' and members of households. The daily 'struggle' of each worker is indeed an investment in her or his household's economic independence and 'self-sufficiency' (a notion carefully examined in the ethnography that follows). The concluding section of the chapter further elucidates the agonistic or 'struggling' aspect of work in Vassilikos. Manual labour in the cultivated fields and around tourist enterprises is perceived by the local farmers and tourist entrepreneurs as a 'struggle' (*agonas*) indicative of a contest between any given human actor and the surrounding environment or 'nature' (*fysi*). As such it directly informs the Vassilikiots' relationship with the physical environment, which is the central theme under investigation in this book.

Investing work in both tourism and farming

A concern about the future of agriculture in the modern era of tourist development is introduced by a claim articulated by the older Vassilikiots, namely that 'the young people have abandoned the cultivation of the land and are solely preoccupied with the business of tourism'. Admittedly, these statements reflect the transition from an exclusive reliance on subsistence farming, to a new situation where tourism-related enterprises provide the greater part of people's income. For the older men and women, who spent the early part of their life working the land and utilising any available resource provided by it, the new generation of Vassilikiots, who often neglect the fields they inherited 'from their fathers', appears as somewhat 'sluggish', or at least 'unappreciative'.

However, these pessimistic statements, expressed predominantly by elderly Vassilikiots, do not accurately portray the observable economic reality of the community. The transition to an economy that is not solely dependent on farming has not imposed a complete abandonment of agriculture and animal husbandry. On the contrary, most of the economically active individuals in Vassilikos continue to be involved in traditional farming jobs, especially when they feel that a decent profit can be made out of them. Unlike their parents and grandparents, they have a greater choice of cultivation options, and feel free to prioritise jobs which guarantee a sufficient profit for the minimum of invested labour. Their more relaxed attitude towards agriculture contrasts sharply with their forefathers' devotion to it. In other words, self-sufficiency, as an ideal code enforcing the maximisation of all subsistence resources that one's land can provide (cf. du Boulay 1974: 244, 247; Kenna 1976b: 349–50, 1990: 151–2, 1995: 135, 2001: 32; Just 2000: 203), does not exert the same kind of pressure on the younger generations of Vassilikiots.

It is hard to attain a clear divide between the representatives of the younger and older generations of Vassilikiots. Most of the forty, fifty and sixty year old men and women participate dynamically in a wide variety of agricultural tasks. Some are successful in recruiting their sons', or indeed their daughters' labour, others are not. But the tension arising from such disagreements is not particularly serious, especially when the sons or daughters have already successfully entered the sector of tourist-related enterprises. When put in this perspective, the complaints of the older folks about the 'young people's neglect of the land' are better understood. A sixty-year-old local man elaborated on this:

> Look at my vineyard. My son, although he learned the skill from me, does not do much work on it. Dionysis had the best vineyard in the area, but he got older, and the vineyard was lost because his son is useless (*akamatis*). But my son, unlike Kostas's son, gets good wages in tourism (*kanei kalo merokamato me ton tourismo*). I cannot press him to do more with the vineyard…

It must be noted that vine cultivation in Vassilikos is not intended for commercial profit. In addition, it requires significant labour. This is why some of the existing vineyards are neglected by the younger men who do not have enough incentives to perform the annual 'pruning', 'cleaning' and 'weed-removal' that a vineyard requires. But this is not always the case. For example, the son of the old man whom I quoted above, is retaining his vineyard, although he is not doing as 'much work on it' as his father expects. His vineyard is small, like all other vineyards in Vassilikos, but for a restaurant owner like him, producing some wine of his own appears to be an additional benefit in the social arena. As there are as many as forty *tavernas* or restaurants in Vassilikos, the aura of tradition associated with locally produced wine, is an extra incentive for the younger, tourism-oriented Vassilikiots to engage in some viticulture. This is an example where tourism reinforces agricultural folklore, adding new value to older traditional motives, such as the cultural significance of home-made wine consumption (cf. Gefou-Madianou 1992b).

In the midst of the summer most Vassilikiots work frenetically to respond to the demands of the tourist economy. This does not mean that during that period cultivation and farming duties are completely neglected or abandoned. Although the tourist economy thrives during the summer months, the inhabitants of Vassilikos do not sever their relationship with the land and agriculture. The harvesting of 'salad' vegetables, such as tomatoes or cucumbers, and summer fruits, such as melons and watermelons, coincides with the massive influx of tourists, and the produce is readily appropriated for the local demand. In fact, the tourists pay well for local varieties of fruit and vegetables, which are available at the local mini-markets and general shops. Vassilikiots, on the other hand, proudly display locally produced products in their shops. They are even prouder when they use those products for cooking in their *tavernas*. 'We feel that we offer the tourists a good service', explained a Vassilikiot woman deeply immersed in the preparations for the seasonal reopening of her restaurant, 'they eat food produced on our land, our own (*ta dhika mas*) vegetables, our own eggs, our own fruits.'

Likewise, chickens and rabbits from the Vassilikiot farms are frequently consumed in the local *tavernas*. In some cases the same household that owns the farm, also owns the *taverna*, and the same men and women who had previously 'taken care' (*eikhan frontisei*) of the wellbeing of the animals to be consumed, have later to kill the animals and help in the cooking. The *taverna* owners who can claim that they are in a position to serve their own (*to dhiko tous*) locally produced meat take a very special pride in this and make sure that their customers – especially the Greek-speaking ones – are aware of the local specialities on offer. 'There is nothing tastier than our Zakynthian *stifadho* [a recipe for cooking rabbits or hares]', some Vassilikiot *taverna* owners boast, 'especially when it is made with our own rabbits (*ta dhika mas kounelia*)!'

In the early autumn agricultural activities regain part of their significance in terms of locally expressed concerns and priorities. Most Vassilikiots do not hesitate to express their exhaustion after the stress of the summer months and welcome the slower pace of traditional farming jobs. But work investments in tourism are not discontinued. Some Vassilikiots devote time to preparing new summer tourist-enterprises, others do not forget to renovate the older ones. Facilities and equipment 'for rent' require constant maintenance work, and areas adjacent to tourist enterprises need clearance from weeds and other kinds of unwelcome flora. This latter task involves a constant struggle with the regenerative power of wild flora, a confrontation for which – as I shall further explain in the following sections – Vassilikiots are equipped, as farmers, with a great deal of skill and patience.

Farming in Vassilikos, although less profitable than tourism, when practised at the right time, with tangible objectives and targeted at the substantial, local tourist market can potentially provide the cultivators with a significant income. In addition to the direct profit derived from marketing the farm produce, the engagement with cultivation and animal husbandry is further rewarded with some additional economic incentives. For taxation purposes, the vast majority of Vassilikiots are registered as 'farmers' (*agrotes*), and receive a considerable amount of state or EU benefits, given to encourage agriculture and animal husbandry. For several Vassilikiots, this is a considerable material motive to perpetuate their involvement in traditional farming activities (cf. Greger 1988). EU benefits, like the profit from tourism, are thus locally perceived as available resources welcomed as credits contributing towards a given household's self-sufficiency. An elderly woman sums it all up:

This is how we keep (*kratoume*) our house(hold)... The subsidy from the EU, the vegetables and the meat from the animals... Even the eggs from our chickens... We also have a couple of rooms 'for tourism' (*gia ton tourismo*). Everything counts... This is how we have made some progress (*prokopi*).[2]

Conflicting discourses on tourism and farming

There are several individuals in Vassilikos who openly declare their preference for farming. As I stated before these are usually the oldest members of the community. They often accuse young men [more often than young women] of neglecting the cultivation of their land. Sometimes, however, anti-tourism sentiments are similarly expressed by young or middle-aged Vassilikiots. 'We are independent (*anexartitoi*) of obligations to other people', they say, after comparing their personal involvement with agriculture, animal husbandry or the building trade with the demands required by tourist enterprises. According to this point of view, tourist entrepreneurs are the 'slaves' (*oi sklavoi*) or 'servants' (*oi ypiretes*) of foreigners, having to 'put up with' (*na anekhontai*) all kinds of eccentricities and satisfy various, unpredictable demands (cf. Boissevain 1996b; Sutton 1998: 26). This is why some local men and women express their antipathy towards the uncomfortable socialisation required by tourism, with comments like: 'We have our land and our animals. We don't have to serve other people.'

Furthermore, Vassilikiots recognise that tourism, although able to provide significant profits in relatively short periods of time, entails elements of uncertainty. It is depended upon 'political, military and economic changes at a global level' (Mitchell 1996: 216), forces and events far beyond Vassilikiots' control. In daily conversation, complaints are frequently raised about the helplessness of the tourist entrepreneurs in controlling the numbers of tourists in their locality. Economic success or failure in any particular tourist season seems to depend on factors external to the local community. This is why several local women and men argue that work invested in farming jobs is 'a security' (*mia sigouria*). Small-scale farming offers the potential for an alternative income, and a strong sense of independence from any uncontrollable external forces affecting the tourist economy (cf. Greenwood 1976).

The local perceptions of tourism, however, are not confined to negative criticism and pessimistic declarations. Several young individuals strongly identify with the role model of the tourist entrepreneur, at the same time as reproaching those among their fellow villagers who retain the lifestyle of the 'old agriculturalist'. The latter were described to me as people who 'spend their life in misery' or 'make their lives miserable'[3], engaged in laborious agricultural jobs that bring little profit. The supporters of a tourist economy point out that even those Vassilikiots who emphatically express their dislike of tourism do eventually engage, to a greater or lesser degree, in various economic activities related to tourism. Their argument is easily verifiable, since almost everybody in Vassilikos enjoys some direct or indirect benefit from tourism. Even the individuals who focus on farming and animal husbandry in the most dedicated fashion enjoy some improvement in their living standard because of the introduction of tourism: better road infrastructure and services available in the community, part-time jobs available for the young members of their households, some cash provided by the direct consumption of their farming products in the tourist market.

Paradoxically, it is a matter of common consensus in Vassilikos, that tourism has benefited (*ofelise*) the community (cf. Stott 1985: 188, 197, 205). Even those Vassilikiots who persistently declare that they prefer farming to tourism admit that, if it were not for tourism, many young people, especially those with insufficient land, would have migrated elsewhere to make a living (cf. Waldren 1996: 230). A sixty-year-old local man illustrates this point:

> I am glad to see young people of our village stay. We had a struggle (*agona*) to come back [from the places we migrated to out of poverty]. Nowadays, Vassilikos is at its best (*stin kalyteri tou*). A little more could be built; but we don't want too much.

Comparisons of this sort, stressing the poverty faced by Vassilikiots in the past compared to the relevant material comfort of the present, are frequently articulated in local conversations. Even the most severe critics of tourism have a few equivocal remarks to make about the recent prosperity their community enjoys. As an old shepherd repeatedly pointed out to me: 'When I was young I had teeth but no food to eat... Nowadays, with tourism (*me ton tourismo*) I have plenty of food to eat, but I have no teeth!'

Evidently, two separate conflicting discourses about tourism and agriculture exist in Vassilikos. The first epitomises the advantages of traditional farming activities and underscores the disadvantages of tourism. The second argues for the reverse; the discomforts of the farming lifestyle are emphasised, while the benefits of tourism are highlighted. Between these two ideological poles – most often represented by older folks who consistently express their nostalgia for the vanishing farming lifestyle and some young individuals who persistently criticise the lifestyle of the old-fashioned agriculturalists – the great majority of Vassilikiot men and women find a voice. They are perfectly capable of contributing to both discourses, at different instances, provoked by different economic or social dynamics in the context of different conversations. A not particularly profitable tourist season, for example, or even various incidents of tourists behaving 'improperly', could instigate a discussion in which the negative aspects of tourism are vividly elaborated and the old farming ideals revered. The same rhetorical fervour is often applied to expressing disappointment in a poor olive harvest or a prolonged drought; but this time it is the misery of the farmer's life which is portrayed and the unrewarding aspects of agricultural labour that are overstated.

The majority of Vassilikiots constantly shift between the two alternating identities of the farmer and the tourist entrepreneur with surprising ease and spontaneity. The tourist economy provides them with exciting financial opportunities; those who own land in the vicinity or have well-established roots in the community are supposed to be the first to legitimately exploit the new resources. However, lack of experience in the new forms of enterprise make most Vassilikiots feel uncomfortable or insecure. When difficulties in the tourist sector arise, they find consolation in the well-established and morally safe identity of the farmer. Agriculture epitomises security in the material sense, while at the same time it provides Vassilikiots with a form of moral and psychological protection, a remedy for the complications lurking behind the precarious constitution of the tourist industry. In the following section, I will focus on the specifics of agriculture in Vassilikos, offering a glimpse of the engagement of some farmers with it.

Notes on cultivation and labour

'The basic products of Zakynthos are oil, wine and raisins; but in Vassilikos basically we do oil.' This is how the older Vassilikiots laconically refer to agricultural production on their land. 'We also used to do wheat and hay straw', they add. Nowadays, unlike earlier times, wheat is rarely cultivated, but some fields are ploughed and sowed to produce fodder. Some of those fields are fenced and flocks of sheep are allowed to enter and graze in the dry season, when food is not available elsewhere. On farmland situated in proximity to domestic units, the villagers cultivate vegetables, including tomatoes, aubergines and beans, in polytunnels or outside in the open fields. Melons and watermelons are cultivated in fields where the soil retains some moisture and does not have to be irrigated. But the majority of the cultivated land in Vassilikos is covered with olive trees. As I will describe in detail in the following chapter, the harvesting of olives is the most intensive economic agricultural activity in the area, and olive oil the most widely and copiously produced agricultural product.

I have already mentioned that Vassilikiots cultivate vegetables in gardens (*mpostania*) located, in most cases, close to their dwellings. Some of them construct polytunnels, which are locally referred to as greenhouses (*thermokipia*). They prepare the polytunnels in early spring, aiming to provide the local market with tomatoes by May or June. The price of the early tomatoes grown in this period is high and the cultivators are usually satisfied with the profit. Later it falls, as tomatoes planted in the open fields enter the market. Other vegetables, like beans, cucumbers and aubergines are cultivated along with tomato plants in the polytunnels or outside. But, during summer time, most of the local cultivators focus primarily on tomato cultivation, a vegetable enjoyed by locals and tourists alike in the form of Greek salads.

Vassilikiots usually produce the seedlings for the tomatoes they cultivate themselves. The seeds, however, are acquired from the Department of Agriculture, and are supposed to be monitored bio-technologically so as to ensure maximum productivity. The farmers in Vassilikos plant the seeds in primary seedbeds, where the tomato seedlings grow unhindered, until they are finally replanted in the polytunnels or out in the open fields. Those seedbeds are covered with transparent polythene sheets. The polytunnels are covered with the same material, and their frame is constructed of reeds and wooden poles, like cloches. Parts of the same material may be used for the construction of a new greenhouse the following year. The ethic of self-sufficiency rules here and the local farmers utilise whatever resource already exists on their farms, buying new materials only when they have no choice. This is what I recorded in my fieldnotes:

Today I was working on one of the local farms. The farmer was constructing a greenhouse, building the frame of it with reeds and wood already available on the farm. As I soon realised the reeds were the ones we had cut together the day before. The farmer gave himself credit for having planted the reeds in a swampy spot on his land, unsuitable for cultivation. He commented upon how, by planting a few roots of reeds on that land he now had access to a useful resource, one that was so plentiful that he could share it with others [in a context of delayed reciprocity and exchange of small favours]. 'At first there were only a few roots,' he said, 'now there are so many that others come and take them.'

Despite the heat, our work steadily progressed, thanks to the farmer's slow but steady tempo. While working, I kept admiring his ingenuity in discovering a new use for old wood and other discarded material. In fact, he took a great deal of pleasure in reinventing a new function for second-hand materials, such as wood, bits of string or polythene sheeting. He shared the excitement of this inventiveness by making frequent comments about the use of various things, jokes and other pointed remarks.

Vassilikiots appear happy with the introduction of chemical fertilisers and new bio-technologically improved varieties of seed to their cultivation. In the context of other discussions, such as those revolving around the decline of hunting prey [see Chapter 7], Vassilikiots may appear sceptical about or, even, critical of the introduction of modern pesticides, which, as they admit, 'poison all the little birds and small animals.' However, when it comes to making strategic decisions about the future of their cultivation, most local farmers acknowledge that pesticides, fertilisers, greenhouses and biotechnology, enhance agricultural production and bring considerable rewards for 'the farmer's struggle and hard labour.' Here is how two Vassilikiot farmers talk about the tomato seedbeds, their greenhouses and the introduction of new seeds and fertilisers:

In my seedbed, I am using seeds from America; they are 'regulated' (*rythmismenoi*) by the Agricultural Control. I was given these seeds by the Agricultural Co-operative (*synetairismos*) in the town. The soil I am using for the seedbed is 'special' (*eidhiko*), 'with vitamins and trace elements (*stoikheia*)'. Not like the old times when people had to weed all the time (*na xekhortariazoun oli tin ora*)!

It is thirty years now, since we started using greenhouses. They were first used in Crete. In the old days we made [selected] the seed ourselves. We had tomatoes only in their normal season. So, we used to cut them into halves, dry them in the sun and put salt on them. In this way, we had tomatoes for cooking during the winter. Since we started building polytunnels, we have fresh tomatoes most of the time! As you can see the production is good, thanks to the new improved seeds and the fertilisers.

The soil in some fields in Vassilikos is very suitable for melon and watermelon cultivation. In those fields a local variety of melons, the 'Zakynthian water melon' was cultivated in the past, but not any more. The cultivators do not regret the loss of the local variety of watermelons. They argue that, 'those melons were tasteless and they didn't bring in a profit'. This is why they replaced them with smaller varieties, those that are popular with the tourists. Their long experience in watermelon cultivation, however, helps them identify the most ideal parts of their land for planting the watermelon seedlings. Under the dry surface they detect plots of land where the soil retains enough moisture to sustain a whole yield without the need for frequent watering. The watermelons produced, although small in size, are extremely delicious.

Unlike work in the greenhouses – where the hot temperature dictates a slow pace of work – the work in the open fields, where melons and watermelons are planted, is more intensive. I still remember my exhaustion on a hot sunny day in early May, when I was helping in the planting of melons along with two senior Vassilikiots. We had to dig holes and bend down to plant the melon seedlings into the soil. Then we carried water in big buckets for some considerable distance, to water – for the first and probably last time – the seedlings already planted. But the stamina of the two sixty-year-old farmers I was working with was remarkable. They often had cramps in their legs from bending down, and they frequently complained of the hot sun. But the complaints were expressed in a cheerful manner. The sun was personified, and their old age was treated as a topic of good-humoured self-ridicule: 'Old man, you've forgotten how to do the job, and the sun is laughing at you!' One of the men was wage labouring for the other. The latter was careful to communicate his remarks indirectly, through jokes (*bartzoletes*), out of respect for their long friendship and the labourer's age. As for myself, as 'the young lad', obviously exhausted by the hardship of manual labour but too proud to appear weaker than the older men, I was consoled by an abundance of ethnographic riches in the form of jokes exchanged and other pointed comments. Ultimately, I was promised a taste of the melons as a reward for my labour!

Work as 'struggle'

...the winning of bread from the rocky fields is, as the villagers say, 'an agonising struggle' (*agonia*). For the greater part of the year nature, if not actually hostile to man, is at least relatively intractable. Day after day the farmer wears himself out in clearing,

burning, ploughing, double-ploughing, sowing, hoeing, weeding; all through the year
there are risks from hail, floods, drought, locusts, diseases... (du Boulay 1974: 56).

Work in the fields of Vassilikos is a constant process of investing labour in the
land through cultivation. But Vassilikiots rarely refer to the term 'cultivation'
(*kalliergeia*). They prefer to use the word 'work' (*dhouleia*), instead. Work is often
synonymous with the image of manual labour, toil and bodily sweat. During my
desperate attempts to participate in cultivating the fields, I often encountered Vassilikiot men and women on the village main road, who on noting my mud- or
dust- covered clothes would utter one interrogatory word: '*dhouleves?* (were you
working?).' According to their notion of 'work' as physical toil, 'writing a book
about the village' – my self-presentation as an anthropologist – did not involve
enough physical effort to be considered as proper 'work' (cf. Danforth 1989: 39).
White-collar occupations, although referred to by Vassilikiots as 'jobs' (*dhouleies*),
the same term as 'work' (*dhouleia*), are deprived of the aura of real manual labour
in the fields. This does not mean that white-collar jobs are perceived as inferior
to agriculture work. On the contrary, they are judged to be more comfortable and
privileged occupations, associated with status and financial security. But there is
something special about manual labour, a quality of striving and endurance,
which is met with silent respect and appreciation by most people in Vassilikos.[4]

This highly appreciated quality of 'work' is not merely associated with the symbolic attributes of working the land, but it is extended to any kind of task which
involves physical toil, like 'building work', shepherding, working in a *taverna* or
other tourist related enterprise. It is better described by the word 'struggle'
(*agonas, pali*), a term acknowledged by anthropologists who have studied rural
communities in Greece and Cyprus (Friedl 1962: 75; du Boulay 1974: 56; Kenna
1990: 149–50; Hart 1992: 65; Dubisch 1995a: 215; Argyrou 1997: 163). The
farmers in Vassilikos refer to their work in the fields, or to any other activity
which is physically exhausting, as 'struggle'. They will typically reply to the question: 'How are you?', with the stereotypical expressions: 'We are struggling
(*palevoume*)' and '[We are engaged] in the struggle (*ston agona*)'. Accordingly, they
see the process of cultivating the land, or any other manual work on their farms,
as a process of struggle, a contest with the physical limits of both the labourer's
body and the environment. As Argyrou (1997) puts it, this is,

'a struggle that is carried out at the most basic level of existence. At this level the world
must be dealt with physically; it must be made to submit through the expenditure of
muscle power and sheer determination (1997: 163).'

Farming work in Vassilikos is action that involves 'struggle'. It is a matter of
observation and experience for any farmer to realise that manual labour and effort
is needed in order for the land to become fruitful and its productivity fully
realised. This empirical fact is further acknowledged in the religious cosmology,

with the metaphor of 'Man's fall' and God's imperative: 'you shall gain your bread by the sweat of your brow' (Genesis: 3,19). Like Adam and Eve, Vassilikiot farmers, from the very first instant they acquire land of their own, become engaged in a continuous process of 'struggling' with it. This contest begins with the transformation of bush into cultivated land, and/or 'safeguarding' the cultivated fields from returning to wilderness. The cultivated fields, as part of nature (*fysi*), contain a potential for constant regeneration. They yield vital products under the farmer's care, and weeds – 'the vegetable counterpart of animal pests' (Knight 2000a: 4) – if they are neglected.

Using fire, pruning-shears, scythes and sickles the farmers in Vassilikos constantly try to keep undesirable vegetation under control. They have to struggle, in ditches close to their homes, in the fields, or even on land adjacent to their tourist enterprises. Modern chain-saws or other mechanical devices for pruning are sometimes used in this process, but most often Vassilikiot men and women control the 'wild vegetation' (*agriadha*) with the highly traditional equipment mentioned above; and this requires a lot of hard manual labour. In fact, controlling wild vegetation well deserves to be accounted a 'struggle', since most of the weeds or thorn bushes exhibit a remarkable ability to resist extermination: they prick, have hard stems and roots, multiply and grow rapidly.

In the battlefield of weed and scrub clearance particular plant species taken on personalities of their own – some grow faster or are more prickly than others – and Vassilikiots are often tempted to curse them, swear at them, and address them directly with complaints. 'I cut you down and then I cut again… But there you are, damned thing…,' they exclaim while wiping away their sweat, 'how many times do I have to pull out your roots?' Persistence is not merely the attribute of the wild plants; it is also a virtue of the human protagonists. They never stop fighting undesirable vegetation, often having to clear the same plot of land time after time within short periods of time. The memory of a farmer's voice is still in my mind:

> Look at those weeds (*paliokhorta*)…look at them! They come out of nowhere…it was only a month ago when I cleared [destroyed] them (*otan ta khalasa*). Here they are! They never stop growing… This struggle will never end (*aftos o agonas dhen tha teleiosei pote*)!

This repetitive and constant confrontation with wild flora is not merely undertaken in the context of farming. Far beyond the olive groves and vegetable gardens, the 'struggle' of fighting weeds is carried out in the yards of the local *tavernas* or on the land surrounding 'room-for-rent' establishments. In the land next to tourism facilities the pressure of keeping the vegetation under control is much higher. The local farmers, who are simultaneously tourist entrepreneurs, apply their farming standard of what constitutes a properly managed environment to their tourism-related activities. According to their aesthetic criteria, 'a well-cared-for piece of land' is one subjected to human labour and control to such a degree that the human input or 'sweat' (*idhrotas*) is easily perceived and appreciated by guests and tourists alike.

Another way of controlling wild vegetation is ploughing the fields with tractors. Ploughing is done with the intention 'of clearing the land' and enhancing the productivity of olive trees and other fruit-bearing trees. Vassilikiots claim that 'their lives were eased' by the introduction of mechanical ploughs in the 1960s.[5] Ploughing the fields with cattle, horses or donkeys, a job traditionally performed by men, involved considerable physical effort or 'struggle'. The same was true for the task of harrowing the soil, a job usually performed by women. Nowadays, the tractors plough the ground around the olive trees at least twice a year, and the farmers seem content with the efficiency and speed of the process, as well as the aesthetic appearance of their well-ploughed farmland. 'Look how it looks now!' they say with pride and contentment, 'the wild-vegetation (*agriadhes*) is gone, and the whole has become became more tamed (*imerepse*)!' Scattered in the olive groves lie small tourist apartments, which now, after the ploughing has taken place, become more visible from the main village road. Their owners take pride in pointing to them and remarking in poetic and self-reflective mood: 'This is the fruit of our struggle'.

There is an additional, archetypal form of 'struggle' that directly concerns both tourism and farming. This is the battle (*o agonas*) with the climate. During prolonged droughts Vassilikiots become anxious about the yields of their fields, or the pasture for their animals. Often their 'anxiety' (*anisykhia*) reaches the point of generalised pessimism, a deep disappointment with their life and the quality of their work. They feel that their labour is 'wasted' (*paei khamenos*) or 'lost' (*khanetai*), and their low morale weakens their desire to struggle. As an elderly Vassilikiot vividly explained: 'If it gives back, you work hard and do not feel it'.[6] Strong storms or winds 'do damage' (*kanoun zimia*) to the greenhouses and the gardens, but most often it is the lack of rain, the most undesirable kind of weather that Vassilikiots complain of. Here is how a forty-five-year old local man dramatically articulates his own 'struggle' with the wind and the drought:

> Get angry my '*palikari*' [brave youth, here he refers to the wind], take everything with you and blow yourself out and relieve the pressure. Blow, blow![7]
>
> Will it be rain again, or not? ...the olives will be lost...everything will wither... Lemons? What lemons? The lemon trees have dried up...the olives... Look at the olives...[to my amateur eye the olives like the lemons were just a little bit thinner than usual!].

Similar attitudes towards the weather are expressed in respect of tourism. But here it is sunny and dry weather, which is praised and treated as desirable. Prolonged periods of cloudy weather (*synnefia*) in the midst of the tourist season become timely topics of conversation in the local general stores which also serve as coffee-houses of a sort. 'All was well at the beginning [of the tourist season], but then, the weather spoilt our plans (*mas ta khalase*)' argue some of the local farmers, who are also tourist entrepreneurs. These are the same people who complain about the effects of drought on the olive production, and fervently long for some rain to come – but yet not too much, since tourists are still around – towards the end of the summer period when the olives start to plump up. When clouds and winds spoil the holiday atmosphere without producing any rain, Vassilikiots become doubly agitated:

> ...it cannot be worse than that... This weather is bad for the olives *and* tourism. Yes, it is very bad for both... This wind will destroy the produce and send the tourist (*ton tourista*) [in generalising singular] away...the weather will be laughing at our fate!

The frequent personalisation of the weather [*o kairos*: a masculine term], is consciously enacted in a waggish, ironical spirit. Vassilikiots enjoy dramatising familiar situations and take pride in articulating their thoughts in expressive gestures and pointed remarks. By theatrically attributing some agency to the weather they make the most of their confrontation with the elements.

'All kinds of work have struggle', Vassilikiot men and women repeatedly stress, 'the olive trees and the farm animals and the *tavernas*'. Work is repetitive, physically exhausting, and, in most cases, continuous. It certainly deserves to be called a struggle. Some parts of the Vassilikiots' daily work, like their engagement with some types of farming jobs, remain constant; others are new, like the booming enterprises of tourism. But all types of work require constant responsibility and care. The element of struggle is related to this constant effort of the Vassilikiots to invest in new and old productive activities. It is a battle with forces external to the self (the natural world, the fluctuations of the tourism economy); it is also a contest – a personal challenge – with the limits of one's self. 'Life is struggle', the Vassilikiots recapitulate, 'life is work (*dhouleia*) and sweat (*idhrotas*)'.

Conclusion

Contrary to the frequently articulated opinion that tourism facilitated the abandonment of agriculture in Vassilikos, the overwhelming majority of Vassilikiots continue to engage in traditional farming practices of one sort or another. In fact, the older residents of the community appear reluctant to abandon the 'farming way of life'. Despite their success with the tourist economy, almost all of them define themselves as 'farmers' (*agrotes*). Utilising any productive resource their farm-land can provide, is for them an imperative, an ideal towards independence from market forces, described by anthropologists as 'self-sufficiency' (du Boulay 1974: 244, 247; Loizos 1975: 41, 50; Kenna 1976b: 349–50, 1990: 151–2, 1995: 135, 2001: 32; Just 2000: 203; see also, Gudeman & Rivera 1990: 44–5), 'autarky' (Stewart 1991: 60, 65) or 'subsistent prototype' (Pina-Cabral 1986:

33)[8]. According to this logic, the benefits of both tourism and agriculture are understood as available resources, arising out of the relationship with the land of Vassilikos. As resources of that kind, opportunities arising out of both tourism and agriculture, Vassilikiot farmers maintain, should not be wasted.

Most of the younger Vassilikiots – though their involvement with agriculture seems more opportunistic than their fathers' – have realised that tourism and the farming lifestyle are not necessarily antagonistic. Tourism provides a direct market for locally produced agricultural goods, while at the same time the folkloric aura of traditional agriculture revitalises tourism through 'staged' – that is, deliberately set up – images of authenticity (MacCannel 1976: 91–107) or in-authenticity[9] (Urry 1990: 11; Selwyn 1996b: 28). It is not surprising, that while the 'old folks' continue to believe that tourists come to Vassilikos attracted solely by 'sun, sand, and sea' (Boissevain 1996b: 3; Mitchell 1996: 203–4), their sons and daughters rediscover old agricultural implements, like ploughs or millstones, to decorate their bars and *tavernas* (cf. Waldren 1996: xviii). As I have already mentioned in the previous chapter, owning, but also working, the land validates an individual's claim to local identity and any rights – like the right to enter the tourist economy – directly related to it.

In daily conversation, Vassilikiot men and women, do not radically distinguish between the labour they invest in agriculture and tourism. They frequently use resources and produce derived from their farms to sustain their tourist-related enterprises and vice versa. Furthermore, they interpret work invested in both economies as an investment in their household's economic and social wellbeing (*prokopi*). Made confident by their engagement in various tourism enterprises, several Vassilikiots take advantage of any resource or benefit arising out of traditional farming: EU subsidies for small-scale animal husbandry or even the direct absorption of local vegetable products by the tourist market (cf. Greenwood 1976: 16). In a similar way, the local farmers' working engagement with the land is locally interpreted as a form of security (*sigouria*), a solid foundation of confidence against the fluctuations and relative instability of the tourist industry or even the disappointment of occasional unrewarding 'dealings' with some tourists (Boissevain 1996b). Although less profitable than tourism, farming entails an occupational identity that the Vassilikiots handle well. This identity is an additional asset in those cases or contexts where tourism 'revalidates' local practices or brings about a 'revival' of interest in local culture (Abram 1997: 46; Waldren 1997: 53).

Finally, with respect to the main topic of this book, the Vassilikiots' relationship with the natural world, 'work in the fields' (*i dhouleia sta khorafia*) and 'work in tourism' (*i dhouleia ston tourismo*) constitute fine examples of the indigenous agonistic and combative attitude towards life. The image of work as a 'struggle' is more than a mere metaphor to the Vassilikiots. It embodies their continual, persistent effort to induce the productive resources of their land – agriculture, but also tourism – to yield the fruit of their own labours. As I will further illustrate in the following chapters, success in farming involves constant care of the land, the continual 'struggle' to keep the disorderly and unpredictable elements of the natural environment in order. In addition, small-scale tourist enterprises, like the ones run by Vassilikiots, presuppose constant care of the necessary facilities and the surrounding environment, a kind of work or 'struggle' very similar to the manual labour devoted to farming. As several Vassilikiot men and women clearly describe, labouring for tourism and labouring for the farm are processes that cannot always be radically separated:

> 'Having rooms for rent in an olive grove requires both the rooms and the olive trees to be well-cared for. You have to work constantly on your land… You have to struggle (*na agonizesai*)… Even tourist jobs involve toil!'

NOTES

1   Leontidou (1995: 97), in her analysis of the effects of tourism in Greece, points out that 'in the initial stages, mass tourism may create jobs and an intensification of agriculture.' 'Later, however, negative consequences are felt, by overcommitments of resources to tourism' she adds (ibid.: 97). In Vassilikos tourism is not developed on a grant scale and although Vassilikiots are often critical of the new tourism economy, they still enjoy the benefits of their recent initiation.

2   Or, as one of Jane Cowan's informants puts it: 'A little from here, a little from there, we will manage' (Cowan 1990: 39).

3   '*Khanontai mes tin mizenia*', '*Kanoun tin zoi tous mizeri*'

4   'Only countrymen know what it is to work…city people…have no idea what it is to work' maintain Lison-Tolosana's (1966: 16) Aragonese informants, echoing the words of my friends in Vassilikos.

5   Here are the comments of a couple of Vassilikiot farmers on ploughing and agricultural machinery:
    'In the 1950s the wooden plough was still used in Vassilikos. The iron ploughs came into the village a few years later. Stelios was ploughing with a wooden one until the 1970s. The iron plough was expensive and he was poor. He still has one at his place. Lefteris, your friend, knows how to make them. That was the job of his father: he used to make things of this kind…'
    'Tractors appeared in the village in the 1960s and after. In 1953 the first threshing-machine came to Porto-Roma [a place-name in Vassilikos]. Now, life is much easier with those machines. But I still harvest a tough piece of my land by hand.'

6   '*Ama apodhidhei, dhouleveis kai dhen to katalavaineis*'

7   '*Thymose palikari mou, parta ola na xethymaneis, na ektonotheis. Fysa, Fysa!*'

8   With reference to 'the conception that a household survives by its own means', Pina-Cabral (1986) employs the term 'subsistence prototype', which is extended to account for a range of local 'images' related to the welfare and reproduction of the household or even to the reciprocity and equality among different households. Although, the term 'subsistence prototype', as defined by Pina-Cabral, appears more efficient in accounting for reciprocity between neighbouring household units, I prefer to refer to the ideal of 'self-sufficiency', because of its more restricted, but more meaningful associations.

9   The recent anthropological and sociological scholarship on tourism acknowledges not merely the tourist quest for the authentic (MacCannel 1976) but also some tourists' willing and conscious acceptance of what they know is not authentic (cf. Urry 1990, Boissevain 1996b, Selwyn 1996b).

5
GENDERED LABOUR IN THE OLIVE HARVEST

Olive cultivation has a long history in Zakynthos and Vassilikos. The rich culture associated with it includes words and terms indicative of the specifics of cultivation, material objects or equipment used, specific roles assigned to the cultivators and harvesters, stories and memories, the cumulative experiences evocative of local social and economic life. The lack of mechanisation of the harvest contributes to the image of olive cultivation as a purely 'agricultural' form of work (cf. Brandes 1980: 138–9; Gilmore 1980: 42), a realm of the Vassilikiots' life which is still relatively independent of the tourist economy.[1] Olive cultivation in Vassilikos, although less profitable than tourism, still attracts the interest of most Vassilikiots and, during the olive harvest period, unites the efforts of individual household members in collective undertakings. In addition, the olive harvest involves manual work enacted by both women and men according to a clear, gendered division of labour, which is part of an olive cultivation culture with deep roots in the past.

This chapter primarily concentrates on the olive harvest, a significant part of Vassilikiots' interaction with their immediate environment and its productive resources, during which both women and men perform hard manual labour. I will approach the harvest as a good opportunity to shed some light on the distinctive logic of this labour – the 'mode of thought that works by making explicit the work of thought' (Bourdieu 1990: 91) – for the actors concerned. Since, 'practices are the means through and the site in which gender ideas and relations are realised' (Cowan 1990: 16), my second concern will be the meaningfulness of women's involvement in the olive harvest. Prioritising local interpretations, I try to explain why some Vassilikiot women prefer work in the olive harvest to other kinds of household responsibilities. I approach Vassilikiot women's deliberate engagement in the harvest as 'a purposive activity', a part of their investment 'towards effectiveness in relationships' (Strathern 1988: 164).

In his description of the olive harvest in Monteros/Andalusia, Brandes accounts for the sexually provocative banter during the harvest by seeing it as a form of release valve for otherwise restricted sexual fantasies and concerns (1980: 137–57). Unlike Brandes's more 'psychological interpretation' (Corbin & Corbin 1987: 166), which 'focuses on abundant cases of sexual tension' (Brandes 1992: 27), my ethnographic material from Zakynthos suggests that the most salient meaning in the Vassilikiot harvesting practices is co-operation. In this respect, my analysis of harvesting practices falls into line with a series of anthropological accounts of Greek communities which emphasise the complementarity (symmetrical or asymmetrical) of gender roles (Campbell 1964; Friedl 1967; du Boulay 1974, 1986; Dubisch 1986; Salamone & Stanton 1986; Hirschon 1978, 1983, 1989; Loizos & Papataxiarchis 1991a; Gefou-Madianou 1992b) and women's agency and ability to make critical decisions about their households and ultimately their own lives (Dubisch 1986, 1991, 1993; Hirschon 1989; Galani-Moutafi 1993).

During my fieldwork in Vassilikos I was lucky enough to record intimate conversations with Vassilikiot women and men reflecting on the local economic logic of investing in household relationships. Short extracts of this discourse, authored mainly by women, are presented in the ethnographic sections that follow. First, I devote some attention to the social history of olive cultivation in Vassilikos, and in particular to the local system of exploitative tenancy agreements that regulated the allocation of the olive produce between landless labourers and wealthy landlords. The remaining sections are a portrait of women and men working and talking in the olive harvest. The particulars of the harvest, and more importantly, the local meaning attributed to the specific form of gender division of labour are thoroughly discussed. I conclude by arguing that Vassilikiot women's engagement in the olive harvest and their decision to retain key positions in collective economic enterprises cannot be entirely separated from the achievements and prestige of their household. Through participating in the harvest, local women prioritise the economic wellbeing of their households and acquire a central role in the management of family affairs.

Before proceeding with the presentation of the ethnographic material, I should like to emphasise that I do not want to suggest that gender relations in Vassilikos epitomise a state of everlasting harmony. Both local women and men make their own choices regarding their investment in common, household enterprises; women, however, may have less options than men. The aim of this chapter is to offer some further insights regarding the meaning of agricultural labour in Vassilikos, prioritising this time some points of view expressed primarily by women. I also take the opportunity to examine the gendered dimension of work in the fields of Vassilikos as one additional aspect of the pragmatic, confrontational relationship of Vassilikiots with the productive resources of their environment.

Tenancy disagreements

An Austrian traveller, the Archduke Ludwig Salvator, who visited Zakynthos in 1901 and 1902, published in 1904 a remarkably detailed account of various aspects of the island's folklore and economic life. Vassilikiots recall stories they heard from their fathers and grandfathers about 'this foreign aristocrat' (*xeno aristokrati*), whom they describe as 'wandering around the island, drawing pictures of houses and landscapes...' (cf. Waldren 1996: 18). This is a brief taste of what Salvator recorded about olive cultivation:

> In Zante [: Zakynthos] there exist several kinds of olive-trees. There are the renowned *dopies* (local) olive trees, which become black very quickly and the well-known *koroneikes*, which come from Koroni and remain green for a long time. Both these kinds of olives are used to make olive oil.
>
> The harvesting of the olives starts in mid October. At this time the locals start beating the leaves with sticks, while a few men use ladders to reach all the branches, even the higher ones. They spread large sheets of hessian on the ground and then they gather the olives in big sacks, which are transported to the olive-mill by cart...
>
> The harvesting of the olives starts after the estimates or *stimes* [evaluations of the produce] have taken place. Those olives that fall on the ground before the estimates, belong to the tenant labourer or to anybody. After the estimates, the local people begin to harvest the olives. The people who do the estimating are called *stimadhoroi*... (Salvator 1904: 470).[2]

Since Salvator's visit, at the beginning of the century, the basic principles of olive harvesting in Vassilikos have remained the same. Although the villagers use tractors for the transport of the sacks, the method of harvesting by using sticks and olive-sheets still prevails, as will be further illustrated in the next section. Until twenty years ago, *stimes* or 'estimates' of the produce in those olive fields which were cultivated and harvested by tenant farmers were commonplace in Vassilikos and even nowadays are not completely abandoned. The kind of olive trees found in Vassilikos continue to be the two varieties described by Salvator. The younger trees belong almost exclusively to the *koroneikes* variety (which is widespread in southern Greece), but the local cultivators still point to some fields with huge, old olive trees of the *dopies* [local] variety and say: 'These trees are very old. They have been here since the time of the Venetians.'

As Vassilikiots themselves suggest, the history of olive cultivation in Zakynthos dates back to the Venetian occupation (1485–1797). The Venetian Democracy demonstrated a fervent desire to encourage olive cultivation (Couroucli 1985: 35,95–8; Hannell 1989: 117), offering a small reward for any tree planted (Salvator 1904). The old olive trees planted in the 'times of the Venetians', explain the farmers in Vassilikos, are arranged uniformly, in parallel lines and at wide intervals from each other. In contrast, olive trees planted in more recent times, are positioned closer together, so as to save space and intensify production.

The vast majority of olive groves in Vassilikos – at least before the Second World War – were the property of landlords (*afentadhes*) while most of the olive cultivators were in effect landless tenants (*semproi*) who lived and worked on the estates of those local proprietors. The status quo of land ownership changed, however, in the three decades following the war. Most peasant cultivators gradually acquired plots of land of their own and planted olive trees on most of them. Nevertheless, and because land holdings in most cases were not enough to provide them with a living, most of these people continued to cultivate the landlords' fields as well as their own. In fact, it was often the very same cultivators who worked for a landlord as tenants before, who continued being employed in his fields even after their acquisition of some plot of land. The landlord was expected, as a good patron, to allocate the cultivation of a field to the man whose family had traditionally cultivated the field for the last two or three generations. What is known today in Vassilikos as *sempria* (plural, *sempries*) is an example of such an arrangement between a landlord and a tenant labourer. The term *sempria* refers nowadays to an informal tenancy agreement that predicates the terms of any given cultivation.

Particular patterns of *sempria* arrangements were applied to olive cultivation to regulate the terms of the cultivation and the allocation of the produce. In the past, the two most widespread patterns were *tritarikes* and *ana pentis*. When a tenant labourer (*kopiastis*) had a *sempria* arrangement for olive trees of *tritarike*s variety, the family of the labourer was expected to cultivate the field, harvest the olives, and deliver two thirds (67%) of the produce to the landlord. According to this arrangement, the cultivator was entitled to one third (33%) of the produce and this was the reward for the labour spent on cultivation and harvesting. A *sempria* arrangement of the *ana pentis* variety had in general the same requirements, but the percentage of the produce allocated to the labourer was slightly higher. The olives harvested were divided in five parts (*sta pente*), three of which were given to the landlord (60%), and two to the cultivator (40%). Some Vassilikiots explain:

> *Sempries ana pentis* were [given] on mountainous or sloping fields, where harvesting was harder and the yield lower. Most of the olive trees on good fields (*sta kala khorafia*) were [given as] *tritarikes*.

Those two patterns of *sempria* arrangements applied to olive cultivation operated in the past as fixed points of reference, saving the landlords from the uncomfortable task of negotiating and renegotiating the terms for each particular arrangement. In addition, a third party called a *stimadhoros*, which literally means an 'estimator' (*ektimitis*), was involved in any *sempria* arrangement.[3] The job of the *stimadhoros* was to estimate the 'expected' produce of particular olive groves. That was necessary because the productivity of olive trees varies from one year to another, being dependent on the climate and the biological cycle of the trees themselves.[4] The *stimadhoros* was always an outsider. If he had been 'a man from within the village' he would have been suspect to partiality, either favouring the labourer due to kinship connections or the landlord due to obligation. Here are some examples of what Vassilikiots remember about those estimators:

> The *stimadhoros* used to estimate (*stimarize*) the produce of a field. He used to say, for example, 'I work out that this grove makes a hundred *vatselia* (*vatseli*: half a sack). If you made more, that was profit for you. But if you made less... In a season with bad weather you could lose out (*empaines mesa*).

> A *stimadhoros* was also a *geometris* (land-estimator), something like a land surveyor, he could measure and estimate the value of land. Some of them had learned their skill by long years of practice. *Stimadhoroi* were always outsiders. The master and lord of the land used to go along with the *stimadhoros* to the fields, but the *stimadhoros* was the one to make the decision. If the labourer disagreed with the estimate – he could say 'that's not right' (*dhen einai*) [literally: they are not as many as you maintain] – the master could arrange for an observer (*parastatis*) to be present during the harvest.
>
> *Stimadhoros*, you said. Yes, *stimadhoros* and *geometris*; this is what those people were called...[a pause]... A few of them were good, but some were devils...

Most olive cultivators in Vassilikos felt relaxed about the relative impartiality of those estimators. This was because they were able to check the estimate themselves, a skill anyone can acquire by experience. Some of Vassilikiots demonstrated this skill to me. 'This field will make an X number of sacks' they calculated. And their estimate was always highly accurate. In the past, they explained to me, if they were in disagreement with the *stimadhoros*'s estimate, an 'observer' (*parastatis*), who was usually the landlord's overseer, arranged to be present at the harvest. The observer was present to measure the actual number of sacks harvested and to make sure that the distribution of the produce was taking place according to the shares established by the *sempria* arrangement, which was, in most of the cases, two parts for the landlord and one for the labourer. A Vassilikiot man in his seventies remembers:

> In the old days there were overseers. For example, one of them could take a villager to court, as though he had stolen something, although everybody knew that he hadn't. The overseer used to say to the judge: 'Give him a small punishment, I just wanted to scare him'.

...and his wife adds:

> The wives of the two big masters [the masters were brothers] used to sit with their embroidery and their magazines, to keep watch on us. They were constantly repeating: 'Distribute well, distribute well' (*moiraze kala*). They used to say this, even when it was *just* about an extra bucket of olives.

71

The latter respondent refers to conditions of cultivation and harvesting prevailing as recently as the early 1960s. The labourers (*kopiastes*) were constantly reminded of the 'right', 'three to one' ratio of produce distribution. The degree of poverty up to that time was such that even an 'extra bucket of olives' would have made a difference to the cultivators and their usually large families. Despite their difficulties, however, most of the tenant labourers in Vassilikos had a reputation for being 'faithful-to-the-master' (*afentopistoi*); they would never 'cheat', even when there was no one present to observe them. The Vassilikiots' loyalty to the landlords was commented upon and criticised by other Zakynthians living in neighbouring villages, but most frequently by Vassilikiots themselves. Numerous Vassilikiots are able to recall instances of fellow villagers – in most cases they are in a position to state particular names – expressing their 'faithfulness-to-the-master' with words like: 'Cheat on my master! I would rather cut off my hand instead.'

Complete 'faithfulness-to-the-master', however, gradually declined as soon as the landless peasants obtained land of their own and grew gradually more independent of their ex-landlords. In the beginning, they persuaded their landlords to cover the cost of fertilisers or to give them more favourable *sempria* arrangements.[6] Later, the blossoming of the tourist industry, which has continued to grow steadily since the late 1970s, provided the majority of peasant labourers with alternative forms of income. The tourist economy forced the few remaining landlords to lower their expectations considerably. This is the point where the intervention of a *stimadhoros* became redundant. Nowadays, the produce can be divided into equal parts or *misakes* (halves), and in some instances the labourers can achieve even more profitable arrangements. An old man, who has been working for years as landless tenant in Vassilikos, explains:

> *Sempries* of olive fields were never *misakes* (halves). *Misakes* apply nowadays, sometimes. But even now… they are rare. Nowadays, most often they are *ana pentis*.

But a younger man, who is currently actively involved in olive cultivation, makes a different assessment:

> Now, you can find *misakes* olives. Now, you can even find [an arrangement] where you can take as much as sixty percent. Especially in rough places. In rough places, you lose time getting the sheets (*liopana*) set properly and in the long run you harvest less sacks.

During my time in Vassilikos, I noticed several cases of tenant labourers (*kopiastes*) negotiating the working terms of *sempria* arrangements with the landlords. This kind of negotiation was, and still is, a slow process. The tenants are content to achieve minor improvements concerning particular terms for cultivation every two or three years. Sometimes they are willing to 'put up' with a disadvantageous arrangement owing to their obligation to their landlord. A forty-year old man, for example, 'has the *sempria* of an olive grove', which was cultivated by his father before him. He is still cultivating the grove with an *ana pentis sempria* arrangement. The man admits that this percentage is low by today's standards. It happens to be the case, however, that the landlord provides him 'with other benefits' (*alles avantes*) related to pasture for his sheep. 'This is why I still tolerate the *ana pentis* arrangement', he points out, 'but this is going to change soon'.

Similar complaints are expressed by the landlords. A descendant of a family of landlords, for example, always gives his olive grove to be cultivated by people who used to be the tenant labourers of his father and his grandfather. He argues:

> I am giving those olive trees to them as 'halves' (*misakes*), which is supposed to be a good deal for them. The profit is small for me. I just earn enough money to maintain the field. It covers the cost of tractor-ploughing and the necessary fertilisers. This is all. The price of olive oil is too low.

The fall in the price of olive oil and the alternative economic opportunities provided by tourism has made some Vassilikiots reluctant to continue undertaking *sempria* arrangements for olive cultivation with the old, traditional, standard patterns. Anger, at the fall in the price of oil, is repeatedly expressed. 'I will not do it again if the prices are like that; it isn't worth the effort', they argue. But at the end of the day they do harvest the olives. They may even sell the olive oil for more than the basic price, and, the next year, they are ready to renew their *sempria* arrangements.

Some Vassilikiots criticise their fellow villagers for their habitual dependency on *sempria* arrangements. One of them said:

> Nowadays there are some good *sempria* arrangements for the *semproi* of the big landlord. But they are stupid. They gamble their money and never have property of their own. Then they are in need of him again.

The man who made this sharp comment managed to minimise his dependency on the landlords after years of hard-working effort. Others are still undertaking *sempria*-arrangements to supplement the profit they make from tourist enterprises, or the cultivation of their own land. During the long winter period, income-yielding opportunities outside agriculture are rare. The mere existence of olive groves owned by landlords signifies a kind of economic challenge for some Vassilikiot farmers – even though olive cultivation, when compared to tourism, offers little profit for a lot of hard work. But for those local families who faithfully adhere to the ethic of self-sufficiency, any resource provided by the land 'should never be wasted.'

Sempria arrangements have been in the past, and continue to be, disadvantageous. Vassilikiots are conscious of the exploitative conditions of such tenancy agreements but continue to engage in them. Yet despite the low price of olive oil and the high percentage of the produce allocated to the landlord, the cultivators always manage to make some profit. There are two prerequisites for this. The first is the co-operation of the whole family in harvesting. Giving one's labour for the sake of the household is a practice that directly depends upon the traditional perception of the household as a single economic unit sustained by the collective labour of its members. The second is the ideal of self-sufficiency, which requires that cultivation is managed through all available means [see, also, the previous chapter]. Olive groves, where they exist, are a resource, and as such should not be wasted, especially if they can be favourably exploited. Co-operation of household members, the ideal of self-sufficiency and their interrelation are themes that will be further elaborated below.

Dividing the tasks

Despite the fact that olive cultivation is a common agricultural practice across the Mediterranean, the particular conditions of engagement with it and its importance vary considerably even within distinct parts of the same geographic region. The manner in which olive harvesting is carried out in Vassilikos, for instance, bears more similarities with harvesting in Monteros/Andalusia as described by Brandes (1980) than with Episkepsi/Corfu, the neighbouring Ionian island which has a social history akin to that of Zakynthos (Couroucli 1985). At the same time, the communities of Episkepsi and Vassilikos still place great importance on olive cultivation, where the harvesting involves a major mobilisation of the local work-force in both places. In Meganisi, on the other hand – another Ionian island near Lefkada – olive cultivation is now declining and it is the preoccupation of the oldest members of the community (Just 1994: 46–7, 2000: 54), while, in Deia, a Mallorcan community, olive trees, which were once described as the 'wealth of the mountains', 'are now potential building sites' (Waldren 1996: 28, 107–9).

Differences in the manner in which olives are harvested can be also found in ethnographic descriptions, especially with reference to the tools and equipment used[7], the duration of the harvest[8], and the gendered division of labour. Unlike Vassilikos and Andalusia, where the trees are beaten, in Corfu the cultivators simply lay the olive-sheets and wait for the olives to fall off the trees naturally. In the latter case, harvest lasts longer (Couroucli: 1985: 109) and men – who do not exhaust themselves by beating the trees – lay the olive sheets (Couroucli: 1985: 110). In Zakynthos, on the other hand, laying the sheets, along with collecting the fallen olives, are the primary tasks of women.

With reference to the gender division of labour, an 'ideal' code can be identified in Vassilikos, as in several other Greek communities (Friedl 1967: 103–4; Gavrielides 1976: 268; Couroucli 1985: 78–9; Hirschon 1978: 72–3; 1989: 99, 104, 143; Greger 1988: 25–6,34–7; Cowan 1990: 49–51; Gefou-Madianou 1992b: 115,121,124–7; Hart 1992: 243–6; Galani-Moutafi 1993: 253–254). In most cases, women appear to be responsible for the care of small animals and the cultivation of vegetable gardens located near the domestic space. Men, in turn, are liable for the wellbeing of larger animals and shoulder duties performed at a greater distance from the homestead, such as animal husbandry, cultivation, or business matters (cf. Pina-Cabral 1986: 83–4).

Although ethnographically valid in general terms, this generalised picture of gender division of labour should not be adopted lightly and uncritically, as particular tasks and 'ideal' codes vary from region to region and from one community to another. Seremetakis, for instance, accounts for the intense female involvement in tasks elsewhere perceived as male responsibility in the area of Inner Mani in terms of the men's vulnerability to feuds and warfare (1991: 44–5), while Handman observes only minor gender differentiation in the division of agricultural duties in Puri/Pilion (1987: 150–151).

More often than not, ideal codes of gender division of labour are not adhered to as faithfully as one might expect. It is important not to overlook the fact that women are frequently forced to take over, rather successfully, a number of 'male' farming jobs either because of male labour migration or other particular conditions (cf. Pina-Cabral 1986). There are many examples of great industriousness and perseverance among the older Vassilikiot women whose spouses had to seek wage labour away from the village in the 1950s and 1960s. Four Vassilikiot women remember and comment specifically on the subject:

> In the past women used to do all the work. Old Mrs. Popi managed to hold her household together without her husband, and she did well. But if the man does not work, or if the man does not work enough, there is no wealth (*prokopi*) in the household. A woman cannot produce the same results as a man. Look at the olive harvest, for example...
>
> Women could even do ploughing and *panoloi* [beating the olive trees from ladders or while climbing the trees], but only when their husbands were absent. When the men are away, women can do everything. Only shopping in the town and driving tractors was never done by women. But as you see, nowadays women do a lot of shopping in the town. The younger ones even drive cars.
>
> The fact that women had to do a lot of heavy male work has to do with the fact that a man often had to go away to earn day-wages, in order to bring more income to the household.
>
> The heaviest jobs always went to the man (*pigainan ston antra*). Women can dig and thresh and do those jobs well. But in the olive harvest, men do the most difficult part. What the women do, putting the olives in the sacks and moving the sheets around is a hard job as well. But the woman cannot climb the tree and beat the olives unceasingly (*astamatita*).

As the last woman has pointed out, uninterrupted continuity is the prime objective of the olive harvesting team. As the harvesters themselves argue, the division of tasks between men and women serves precisely this principle. 'Men's work emphasises a visible contribution and hard physical labour while women's work stresses flexibility of skills, tasks and time' (Waldren 1996: 109). Men beat the olive trees with sticks unceasingly to make the olives fall on olive sheets previously laid under the trees by the women.[9] They interrupt their beating to help with the laying of the sheets only occasionally. The laying of the sheets is perceived as a 'female' job that requires the help of men only if there are not enough women present to carry out the task efficiently. In turn, since the beating of the olive trees is seen by both men and women as the most important and difficult task, any interruption, be it for the sake of the sheets or any other 'menial' reason, is unanimously proclaimed a 'delay'. Thus, the presence of a sufficient number of women indicates that the men will continue to beat the trees almost ceaselessly, achieving the ideal of a local working party, that is, to harvest as many trees as possible in a given amount of time.

The demanding manual task of beating the olive trees continuously is facilitated by two simple, easy to make instruments (both made by men): the *loros* and the *katsurdheli* (or *katsuridheli*). The latter is a rather short, cleft stick used for beating the trees from close distance. The *loros* is a two- or three-metre-long wand [similar to what (Brandes 1980: 143–144) describes as a *vara*] used to beat the olives to the ground. Both tools are primarily used by men, although women might occasionally use a *katsurdheli* but hardly ever a *loros*. In Vassilikos, the person responsible for beating the olives is called a *tinakhtis* and is always a man. 'A good *tinakhtis* does not do the other kinds of jobs' Vassilikiots assert (*o kalos o tinakhtis dhen kanei alles dhouleies*), proclaiming and verifying the importance and high status of this particular job.

'The other kinds of jobs', then, are left to women and consist mainly of carrying and laying the olive-sheets under the trees, as well as, kneeling down for quite some time 'cleaning' the olives on the sheets. 'Cleaning' consists of removing by hand stones and small pieces of wood (*tsimes*) broken off during the beating process. Women also take care of some larger branches with olives, cut by the men on purpose in order to prune the tree and thus hasten the process of harvesting (cf. Kenna 2001: 49). They remove the olives and throw away the branches, occasionally using the shorter harvesting stick, the *katsurdheli*. When a considerable quantity of olives has been accumulated on a sheet – enough to make the sheets too heavy to be carried around – the women place the olives in baskets and throw them in sacks, a process compendiously described as 'bagging up the olives' (*na sakiasoun tis elies*). One woman holds the mouth of the sack open and another fills up a basket and pours the olives into the sack. Usually the older woman is expected to hold the sack, while a younger one with a stronger back lifts up and empties the basket.

The ideal harvesting team, Vassilikiots maintain, contains at least four men and three women. They explain:

> You need four men and three women. The men do the beating. A good *tinakhtis* does no other kind of job. He goes on beating the olive trees. Then you need two *liopanidhes* and one *katharistria*.

Liopanidhes are the women who set the olive-sheets (*liopana*). The *katharistria* (*katharizo* means 'to clear' in Greek) is a woman who separates the fallen olives from the *tsimes*, that is the small pieces of wood.

Those households that are capable of forming harvesting teams without recruiting additional wage labourers are considered to be the luckiest according to the local criteria of self-sufficiency. However, harvesting teams of the ideal size and gender ratio are rare and many households cannot form an adequate party. Failing to recruit a minimum of four working members, they often have to resort to hiring a wage labourer at a considerable price. If the person required is an experienced *tinakhtis* – usually a fellow villager with tried and tested abilities and guaranteed skill – the wage can be quite high. If the help required is that of a *liopanidha*, a middle-aged woman can perform the 'female' part of the harvesting at a lower wage.[10] The availability of Albanian migrant labourers in the early 1990s solved some of these problems as the Vassilikiots were in a position to add to their harvesting parties one or more men able to carry out efficiently a wide variety of tasks for minimum payment.

Children are rarely present at the olive harvest, because the harvest takes place during school or homework time. Education is highly valued in Vassilikos – as elsewhere in the Greek context (Stewart 1991: 126–7; Faubion 1993: 59–60, 134, 189–90; Argyrou 1996: 35–6; Just 2000: 70–9) – and is prioritised over work in the fields. Most local families foster high aspirations for the education of their children, and of their sons in particular (cf. Goddard 1996: 172). But boys, more often than girls, sometimes make clear to their parents during their high school years that they do not wish to go on to higher education. In this case, involvement in the olive harvest is expected, and in fact encouraged, because the rural household – as du Boulay (1974: 86) puts it – 'cannot afford to carry non-working members, except for the very old, or the ill, and even these do what they can'. This is why in the past, and in accordance with the local ideal of self-sufficiency, children did take part in the olive harvest, performing the simplest secondary jobs. Some children, or even young women, used to collect olives from the ground, the ones which had fallen due to a strong wind, and sell them for a little money. Nowadays, nobody bothers to engage in such a desperate and trivial enterprise! A young married woman remembers:

When we were small kids (*mikra*), my brother and I used to gather olives from the ground, those struck by the wind. We used to sell them. Nine *drachmas* for a kilo. We used to put earth and fat-olives together [*khontroelies*: not the ones appropriate for making olive oil] to make the bag heavier!

Harvesting, work and conversation

Despite the hard manual labour necessitated by the harvesting process itself, and the serious manner in which this particular agricultural job is undertaken, harvesting remains a sociable project (cf. Brandes 1980; Waldren 1996: 108) in the context of which gossip and social commentary have their own distinct role. Male and female participants in the harvest engage in social or aesthetic 'code switching' and practical information of all sorts is exchanged (Seremetakis 1991: 46). Younger harvesters of the opposite sex, despite strict supervision, may seize the opportunity to form 'well tried and deep attachments' (Greger 1988: 25). This kind of sociality, but in its most exaggerated form, is reflected in Brandes's (1980) description of sexually provocative banter during the olive harvest in Monteros, Andalusia. Here I must note, however, that in Vassilikos, perhaps with the exception of some well disguised hints and innuendo, the local behavioural codes that characterise cross-sexual conduct remain strictly adhered to.

Fieldwork time in Zakynthos involved my active participation in the harvesting for two consecutive seasons. Several families with whom I was already well acquainted acknowledged my willingness to help voluntarily and accepted me on their harvesting teams. In fact, after gaining some experience with the particulars of the job I was offered on several occasions day-wages of the lower 'female' rate. Contributing to the harvesting process was in turn an excellent opportunity for relaxed conversations with the harvesters and provided a wealth of ethnographic information. A few instances of my field experience narrated below illustrate what work in the olive harvest feels like and further support my argument that women's involvement in it is intentional and rich in social meaning:

> Today the working team was made up of Dionysis, a sixty-year-old man, his wife, his daughter-in-law and a paid labourer, Spiros. Spiros is an experienced *tinahtis* and is paid 7,000 *drachmas* per day to beat the olive trees. The two women will not let him do 'other kinds of jobs' out of respect for his skill in beating the trees. 'He is one of the best *tinahtes* of the village' they said. 'Furthermore', they explained, 'it is a waste to pay somebody so much money for a trivial task, such as laying down the sheets.'
>
> The two women try to work as much as possible, even the older one, the wife of Dionysis. A strong work ethic predominates. The younger woman, Tasia, is worried about her toddler son who is asleep at their house nearby. She knows that, should the child wake up, then her mother-in-law will go to 'care for him'. Tasia, being much younger than her mother-in-law, prefers to stay with the harvesting team and 'work'.
>
> Most of the time they talk while they are working. Sometimes there is a short silence and one can hear the overwhelming sound of the olives falling, like rain, on the sheets, under the rhythmic beating of the sticks. The topics of conversation are various but most concern recent local news.
>
> Today for example, they were talking about 'a good bride', lost by one local man owing to 'the stupidity of his head'. She is now getting married to a man from the town, who has a job in the civil service. 'They will have a comfortable life (*tha pernoun zoi kai*

kota)', Dionysis comments. And Spyros adds: 'The father of that girl produced (*evgale*) good girls'. 'They are a good family. Nothing bad (*tipotsi kako*) was ever heard about them', Tasia remarks.

The discussion extends to various local women. They make evaluations about their degree of involvement in the 'work in the fields' (*sta khorafia*). In the context of this discussion, both women and men praise those young women who work in the fields, in traditional peasant jobs. The so-called 'bourgeois' attitude of disparaging manual labour in the fields is seriously criticised. On the contrary, women who work in the fields are considered to have some kind of quality that makes them 'a better kind of person'. Tasia explains that although she has a good excuse for abstaining from the harvest – her toddler son – she does not like to 'sit at home doing nothing, like some other women do'. Her mother-in-law and the two men highly praise her attitude towards work.

Then the discussion shifts to a village road, which is about to be constructed, and the rights of private road usage. People driving on the main village road close by were waving to the working team, greeting the two men or making jokes.

During that time, the ethnographer, who was admittedly a motivated worker, but a relatively inexperienced harvester, was assigned the female role of a *liopanidha*, which involved helping the women with the olive sheets, cleaning the olives and filling of the sacks. As I realised, those 'other kind' of menial jobs were indeed quite tiring. Confined to a low status task, and tired by the work, I became increasingly annoyed by the older men's disparaging attitude towards the 'lowly' role of sheet laying and cleaning the olives. The younger of the two women explained to me:

A good *tinahtis* has nothing to do with the sheets and separating *tsimes*. It is considered to be a skill (*tekhni*: artistry) to beat quickly and well. It is a matter of honour for the tinahtis to do no other jobs. A good *tinahtis* does not deign to get dirty (*dhen katadhekhetai na lerothei*). The lowly jobs – moving the olive-sheets around and separating the *tsimes* – make you dirty, [from the oil]. These are jobs for women and children.

Olive harvesting and the manner in which labour is divided are organised according to two sets of crucial distinctions. First, the jobs considered to require no particular skill – laying the sheets, cleaning the olives and so forth – but which make one dirty, are clearly distinguished from the more honourable and 'difficult' duty of beating the trees. The latter is performed by men and is a task that does not bring the labourer into contact with fallen olives and the ground. The second important distinction is drawn between beating the olive trees from the ground [what is called in the local dialect *hamoloi* (ground-style)] and beating them from a ladder or after climbing into the tree itself. The latter is called *panoloi* (up-style) and is performed primarily by men[12].

These two dimensions of the gender division of labour – *panoloi-hamoloi* and beating/cleaning – are respected by every harvesting team in Vassilikos. Only sheer necessity (usually the lack of enough women in the team) would force a man to interrupt the beating in order to help the women with the laying of the sheets and the cleaning of the olives. The women harvesters, on the other hand, provided that they have already set the sheets and placed the harvested olives in the sacks, can, and often do, take a small harvesting stick (*katsurdheli*) and beat the trees for a while (while standing on the ground). They would avoid climbing a ladder, however, and certainly they would hardly ever climb a tree.

This kind of gender-specific task allocation is dictated by a cultural logic that values and prioritises the uninterrupted and efficient continuation of the harvest. According to some elderly Vassilikiot women, 'this is why women wage labourers are paid less than men. The one who beats the tree is the one who carries the responsibility for the work. If 'he' is slow, everything goes slowly.'[13] Or as a younger woman explained to me:

> In the olive harvest, men do the most difficult part. What the women do – putting the olives in the sack and moving the sheets around – is a hard job as well. But the woman cannot climb the tree and beat the olives without stopping (*astamatita*).

At the time of my involvement in the olive harvest, I was already well acquainted with the Vassilikiots' respect for any kind of work that entailed physical effort or 'struggle' (*agonas*) (see my earlier commentary on the same term). Nevertheless, I was at first extremely reluctant to accept the differentiation between beating the olive trees unceasingly, a task unanimously portrayed as 'struggle', and the so called supportive role of the cleaners and sheet-layers. My own arduous efforts to write a thesis about Vassilikos had been repeatedly treated by the Vassilikiot men, according to the local cultural logic, as less of a 'struggle', a kind of pastime compared to the 'real' work they were doing in the fields. My revenge was a committed effort to uncover what I perceived as Vassilikiot men's manipulation of women and children, categories with which I found it easy to identify both in the realm of discourse and that of practice. Such preconceptions made me a classic victim of an equally classic male bias (Reiter 1975: 13,15; Moore 1988: 2). My own urban background, as well as my theoretical and political leanings, led me to read male dominance into the gender relationships in the field. The ethnographic understanding of the division of labour and my subsequent analysis of the harvesting process were thus hindered by my own ethnocentric expectations.

Vassilikiots claimed that 'men do the most difficult part of the job'. My initial opinion, based on my harvesting experience so far, was different. My original resistance, however, was gradually worn down when I was finally allowed – after spending considerable time in the harvesting fields – to try the male share of the work. My delight at being able to prove that I could beat the trees unceasingly was succeeded by pain from serious blisters on my hands, the result of the friction of the wooden stick on my palms. Waiting for my wounds to heal, I returned to the role of the *liopanida*, 'setting the olive-sheets', and helping the women with all the 'female jobs'. Confined to 'ground work' and unable to fully partake in women's performances of 'lack of performance' (Herzfeld 1991b: 81), I stared at the over-performing senior males with considerable envy and admiration. I was dirty, tired and in pain, but still content, for I was doing a 'real' job, even if only in support-ive role.

Women and household prestige

My ethnographic material and experience in Vassilikos further substantiates Pina-Cabral's observation that women in a rural setting, where 'production is carried out at the level of the household', are invested with more economic power than their bourgeois counterparts (1986: 84–7). Tasia, a young Vassilikiot woman intro-duced in the previous section, claims to prefer the hard manual work involved in the harvesting of olives to the more 'relaxed' – according to her – duties of child-care. Tasia is conscious of the fact that:

> The people here in the village respect working in the fields more than staying at home with the children. They say about women in the town: 'she sits (*kathetai*) at home all day'.

'Sitting' (*sic*) at home and abstaining from manual labour in the fields, as I have already noted, is a practice severely criticised and even mocked by both male and female fellow villages. Conversely, those Vassilikiot women who participate in eco-nomic activities collectively undertaken by their household are praised as 'real' contributors to the prosperity of their family and their spouses are deemed to be 'lucky'. Those who avoid manual work in the fields do not acquire the same admi-ration and eventually lose out in terms of relative power and prestige. As Pina-Cabral has observed, 'the peasant women who adopt urban mannerisms in order to increase their short-term prestige, are in fact abdicating an age-old posi-tion of relative power and independence' (Pina-Cabral 1986: 86). Echoing Pina-Cabral, the majority of Vassilikiot women recognise that avoiding manual labour in the fields or in the family-run tourist enterprises results in confinement to the home, piercing criticism from fellow villagers and, most importantly, seri-ous compromise of their ability to have any economic clout in their household. Here is what a young married woman told me reflecting on this idea:

> I prefer to go to the olive harvest in November or work at our *taverna* in the summer. We have olive trees, we also have a *taverna*. Why should I let others work and stay at home alone, pretending to be a lady?

Vassilikiot women are conscious of the importance of their labour's contribu-tion, which is understood as a form of investment in the household economy, deserving recognition by husbands, fathers, brothers and in-laws. Such claims for recognition are more clearly expressed by women themselves in relation to the labour they have invested in their parental household before marriage. This is how a sixty-year-old woman refers to her labour contribution:

> I did a lot for my father. A lot of hard work. But I was never given as much as I deserved for my dowry. I did all the jobs. I did [harvested] the olives, I did *hamoloi*. On my knees, I was working on my knees... I was hoeing the soil on my knees. This is why my knees don't support me anymore.

And a twenty-five-year old:

> I was working for years for my father. I was working in the restaurant and in the fields. But he doesn't give to me. He always helps my brothers. He doesn't give me or my sisters enough. Now I work on our own [property], and my husband does not refuse me (*dhen mou khalaei khatiri*).

In Vassilikos, where a strong patrilocal influence was, until recently, regulating postmarital residence patterns, this form of resolute identification of a married woman with her husband's household is frequently articulated. In the past, this shift in a woman's loyalties was facilitated by the dispersed pattern of the village's settlement, which inhibited regular communication between a married woman and her consanguines. Nowadays, distance between households is minimised by young women's access to cars or mopeds, while patrilocality is also threatened by local girls who marry outsiders willing to settle in the locality and reside with their wife's relatives. Women, however, still engage in collective economic activities undertaken by their households – work on farm, fields or tourist enterprises – and regard their work as an important indicator of their role in household affairs.

Vassilikiot women's conscious decision to engage in agricultural jobs, and the olive harvesting in particular, directly relates to at least three kinds of symbolic capital. First, given that women who work in the fields are praised while those who do not are sharply and overtly criticised (cf. Cowan 1990: 38), participating in such tasks is first and foremost a matter of individual status and prestige. Taking part in collective household projects, like the olive harvest, undoubtedly strengthens a woman's position in the household and the community. Thus, Vassilikiot younger and older women plan their course of action according to the symbolic advantages that participation in joint family endeavours promises.

Second, women's willingness to contribute to their household's welfare greatly determines a household's locally perceived status and material prosperity. Vassilikiot women maintain that their households' relative prestige is greatly dependent on their commitment to contribute to its economic strategies. The importance of the female contribution to household prosperity is equally recognised by men. In the male world of the coffee-house, proclamations such as 'she holds her household together' or 'it was really her behind this household's prosperity (*prokopi*)' are not uncommon and demonstrate the difference that a strong female presence can make in the advancement of various family affairs.

Third, it is impossible to separate men's status, from women's eagerness to undertake agricultural work and their involvement in it. Ultimately male and female prestige are linked (cf. Friedl 1967), as individual and household prestige merge for the sake of the welfare of the same corporate entity (cf. Dubisch 1986: 27; Salamone & Stanton 1986: 98–99; Hirschon 1989: 141–43; Loizos & Papataxiarchis 1991: 8). This is precisely where the ideal of 'self-sufficiency', the importance of which I have been constantly underlining in this and the previous chapter, meets with its counterpart, the concept of 'self-interest'. 'Self-interest' (*symferon*) in rural Greece, as other ethnographers have demonstrated (du Boulay 1974, Loizos 1975; and for an urban refugee context, Hirschon 1989), very often relates to the advancement of particular households rather than individuals. Far from being solely inspired by an individualistic logic, 'self-interest' promotes a household's 'self-sufficiency' and it is in terms of the latter that the former gains its full meaning and significance.

Women and men in Vassilikos understand that it is the labour of the former that permits the latter to perform their 'beating-the-trees-unceasingly' share of the labour. Without women's precious assistance, without a supportive team of suitable female workers who competently set the olive-sheets, clean and collect the olives and follow the rhythm of the team work, the whole harvesting enterprise would suffer considerable delay and its efficiency would be seriously compromised. In such unfavourable circumstances, not simply individual women and men, but whole households bound together by the same sense of collective self-interest, would lose.

Conclusion

In Vassilikos systematic olive cultivation dates back to the time of the Venetian occupation (1485–1797). In the past, the terms and the particulars of the cultivation were determined by a feudal scheme of unfavourable tenancy agreements between landless cultivators and their landlords. Although, in time, the conditions changed and most Vassilikiots acquired their own land and planted olive groves on it, some local farmers continue to accept disadvantageous tenancy agreements offered by landlords.[14] The recent involvement of most men and women in the profitable tourist economy, has not eradicated the local perception of olive groves and their produce as a valuable resource which 'should never be wasted'. What might seem unprofitable in the context of a wider economy is readily 'put to use' in Vassilikos (Gudeman & Rivera 1990:190–1) through household-oriented practices. Realised thanks to the collective co-operation of all available household members, such economic decisions appear to be meaningful strategic options in rural contexts, where the collective household logic predominates over individual concerns and dictates that all available resources should be exploited to the full.

The preservation in the original form of traditional methods of harvesting the olives makes olive harvesting one of the last forms of local cultivation to remain unaffected by mechanisation and agricultural technology. The gender division of labour enacted in the harvest divides the duties into those performed above ground level (by men) and those carried out on the ground (by women). In turn this kind of task allocation demands the close and committed co-operation of all household members capable of engaging in manual work with the exception of children who are expected to attend school. In the unfortunate event that members of a given household refuse to give their labour, the whole enterprise is jeopardised. The agricultural produce might be lost, or adversely affected by the weather, or paid workers will have to be hired; in all these cases, the result will be collective loss for the household.

The metaphorical and actual separation of the harvesting into work 'above' or 'on the ground' does not simply relate to the obvious pairs of structural oppositions: *above:below, male:female*. Following Bourdieu (1977, 1990), I see the olive harvesting as a 'field of action' which embodies the gendered division of labour as conceptualised and practised in Vassilikos. The 'logic' motivating the actors involved is to finish the work as quickly and efficiently as possible, by working 'unceasingly' (*astamatita*), ideally with minimal (paid) labour input by non-household members. In turn, repetitive enactment of the gendered and social division of labour produces a concrete, lived notion of 'what is work in the olive harvest', while this particular division is itself informed by this very understanding. 'Human labour not only generates and regenerates organic and social being; it is the means whereby human beings create and recreate the intersubjective experience that defines their primary sense of who they are' (Jackson 1998: 16). Collective engagement in the olive harvesting, and close co-operation between Vassilikiot women and men, as in every other collective household performance, entails statements about the protagonists' identities. In this case, it is their subjective position, as members of households, which ultimately calls for their wholehearted participation in every profitable joint enterprise.

Female labour and organisational skills are crucial to the success of any family in the Vassilikiot social arena. What the numerous in-depth discussions with Vassilikiot women helped me realise is the importance attached by the women themselves to being invaluable to one's own household. By participating in the olive harvesting and investing their own manual labour, the women gain the advantage of controlling the productivity of the team and thus the efficiency of the whole project. Seen in this light, their role is not merely a supporting one. They do not simply reinforce (Friedl 1967: 108), but effectively engineer, their husbands' status and prestige as capable and successful workers. Their commitment is locally interpreted as 'overt evidence of their intentions' (Strathern 1988: 164), while their intentions constitute a clear statement about their willingness to participate in and contribute to the household's prosperity. The presence of men (or women) in the harvest is 'necessary to and created by' the presence of women (or men), and 'the necessity is construed as a matter of the difference between them' (Strathern 1987: 295). Ultimately, both sexes work for the household, a symbolic entity, which is represented by both men and women in the eyes of the Vassilikiot community.

To be an active contributor to the household's economy is, in turn, a position of relative prestige and control that most women desire at any cost. For they are deeply aware of the loss of power that bourgeois women experience (cf. Pina-Cabral 1986: 84–7) when they remain, as the women in Vassilikos put it, 'isolated at home, sitting all day and doing nothing apart from caring for the children'. Work on the olive harvest is 'really hard work' the women of Vassilikos maintain, but several of them prioritise harvesting olives over domestic work and childcare. As Maria, one of the women harvesters, explained, staying at home was a possibility that had actually 'crossed her mind'. But, after giving 'a second thought to the matter' she decided that 'simply' staying at home looking after her toddler son would be inappropriate. She said:

> When I was a young girl and I was living with my father, I had to work on the harvest on my father's land. Now, I am a married woman and I have a baby, and I don't have to do this kind of work. Working on the olive harvest is what I really want to do. This is our property and I care for it. I don't like seeing my husband and my father-in-law wasting time setting the olive-sheets. Neither do I want them to hire Albanians and paid workers. Besides I don't enjoy staying at home with the kid doing nothing...

Vassilikiots appreciate the value of women's labour and women's involvement in household-run economic enterprises, be it in the field of agriculture or tourism. Unlike participation in family ventures, however, non-essential female wage labour is still criticised locally (cf. Hirschon 1989:103; Galani-Moutafi 1994: 119). The criteria that determine the 'acceptability' of women's work are concerned with independence and autonomy (cf. Goddard 1996: 119, 137–9), or male, female and household reputations (cf. Galani-Moutafi 1994: 118–20). For example, 'a young wife who is hired for olive picking is considered to be helping her husband to build up family income through her wages in cash or kind', but a senior wife's work 'on strangers' land' reflects her husband's and subsequently her household's inability to measure up to the local standards of success (Kenna 1976a: 23; cf. Loizos 1975: 55).

My analysis of the olive harvest in Vassilikos rested upon the explication of the notions of 'self-interest' and 'self-sufficiency', two principles which, as I have shown in this chapter, can only be realised in a mutually supportive fashion. Just as one cannot separate self-interest from self-sufficiency of the household, the ethnographic evidence suggests that in mainstream Greek communities (especially, but not necessarily, those in the countryside) men's and women's reputations cannot be advanced independently of one another. Just as 'self-interest' (*symferon*) has a 'household' rather than an individualistic orientation, men's or women's achievements cannot be treated separately from the achievements of their respective households (cf. du Boulay 1974: 169–70; Loizos 1975: 66,291; Dubisch 1986: 27; Salamone & Stanton 1986: 98–99; Hirschon 1989: 141–43; Loizos & Papataxiarchis 1991: 8). Both male and female work and its products can, thus, be seen as 'instruments of relations', while every instance of joint labour for the sake of the household's prosperity 'makes visible a relationship' between women (as well as men) and their households (Strathern 1988: 164).

This chapter has made apparent that work in Vassilikos – or to put it in Vassilikiots' words, 'day-to-day struggle' – is invested in relationships. It has also illustrated that Vassilikiots' work or 'struggle' is informed by some household oriented principles. Similar household oriented priorities extend beyond the context of human relationships to encompass attitudes towards the animate and inanimate world that surrounds the human protagonists. The next two chapters carry out an ethnographic investigation of local attitudes towards domestic and wild animals, which, as I shall demonstrate, are formed according to the same concrete and lived understanding of the human social world.

NOTES

1 Unlike the kinds of cultivation examined in the previous chapter, olive oil production and the harvest take place when the tourists are absent and the oil produced is not merely absorbed by the local tourist industry. It is part of a more general, large-scale agricultural production, which is frequently affected by agricultural policies and fluctuations of the national and European markets.

2 This text was translated from German to Greek by Angeliki Apergi and Tasia Kolokotsa, then translated into English by myself.

3 Since the Ionian islands were under Venetian occupation for more than four centuries, a lot of Italian words – especially those related to commerce, law and government – penetrated the local vocabulary and became hellenized by acquiring Greek endings. '*Stima*' (evaluation, estimation) and '*stimaro*' (to evaluate/estimate), come from the Italian terms *stima* and *stimare*.

4 For example, the olive production is always higher in one season and then lower in the next. The alternate harvesting season, with its greater productivity, is called in Zakynthos *ladhia* (cf. Just 2000: 54; Kenna 2001: 49).

5 '*Valtou ligo, na ton tromaxo ithela*'.

6 For example, *ana pentis* [40% of the produce allocated to the labourer] instead of *tritarikes* [33% of the produce allocated to the labourer].

7 On Samos, cane sticks are used to beat the olives off the trees (Galani-Moutafi 1993: 254), while in Zarakas/Peloponnese the trees are 'combed with short rakes' (Hart 1992: 243). Both these approaches to olive harvesting are considered 'harmful for the trees' by the cultivators of Vassilikos.

8 Compare, Zarakas/Peloponnese: late October to Christmas (Hart 1992: 242–3) with Episkepsi/Corfu: November/December to May (Couroucli 1985: 109), and Pouri/Pelion: November to March (Handman 1987: 88).

9 In the past the olive-sheets were made from old pieces of cloth or hessian. Women would frequently repair the sheets since they were not easily replaced at the time. Nowadays, most olive-sheets are made from plastic tarpaulin, are lighter and easier to carry, and are easily available at the market.

10 In 1992 and 1993 the day-wages for an experienced *tinakhtis* were 7000 to 8000 *drachmas* per day (and 4000 to 5000 *drachmas* for a *liopanidha*). It is interesting to compare these figures with the

ones provided by Kenna for the 1966/67 olive harvest on the Aegean island of Anafi: '40 *drach-mas* per day for women harvesters, 50 for men; 30 *drachmas* per day for women olive-pickers, or one *oka* of olive-oil' (1990: 148). With respect to the Ionian island of Maganisi, Just (2000: 61) reports that at the 1970s 'a day's labour picking olives was reckoned at 1000 *drachmas* (plus food).'

11 Salvator records in 1904: 'The olives on the ground fallen after the harvesting has taken place, belong to anyone who happens to pass by and takes them, and those people are usually children or women who gather the olives in their baskets' (Salvator 1904: 470).

12 On gender roles and agricultural tasks, Pina-Cabral notices a differentiation between 'products of the air (things which grow well above the ground level)' and 'products of the ground (things which grow in or near the soil)', the former being the responsibility of men and the latter of women (1986: 83). 'Males look up, females look down' comment the Portuguese farmers, when they collectively participate in agricultural work that involves the co-operation of both sexes (ibid: 84).

13 Equal pay for both sexes was institutionalised by Greek law in 1984 (Tzannatos 1989). This is a well-known fact in Vassilikos, and this is why the Vassilikiot harvesters felt obliged to offer a justification for women's lower wages at the olive harvest.

14 A farmer's willingness to accept an unfavourable *sempria* arrangement may be partially dependent upon a previously established 'obligation', involving various kinds of resources or advantages that the farmer has previously received from the landlord. The farmer, however, will attempt to account for the conditions of this relationship, and the requirements of the particular cultivation, by mobilising the labour of his household, and will eventually incorporate the benefits and the resources received into his/her household's economy. The ideal of self-sufficiency rules here, and determines the farmer's economic strategies.

6
ORDERING ANIMALS ABOUT

The people in Vassilikos maintain that they 'keep' animals 'on their land' because animals are 'useful'. They also say, that they 'keep' animals 'on their property' because 'they always did', that is, because 'they are used to having' animals and 'they like' to do so. But then they conclude that they 'like to have animals because their animals are useful!'

The concept of usefulness is central in most local rationalisations concerning animals and animal husbandry. Some Vassilikiots claim that they prefer to work 'on the' animals (*sta zoa*), rather than working on building construction (*stin oikodhomi*) or 'for the tourists' (*gia tous touristes*); but they immediately hasten to point out that the latter offer better economic rewards, and 'this is why' they often 'have to' give them priority over animal husbandry. As I shall demonstrate in this chapter, Vassilikiots' relationship with their animals has some intrinsic value for them. However, this is never explicitly stated or offered as a justification for their engagement with small-scale, relatively unprofitable, forms of animal husbandry. The farmers of Vassilikos admit that they 'like' or 'love' animals, but after a short silence, they add an explanatory phrase starting with the word 'because': 'because... it is good to have animals', 'because animals are useful'.

'A distinction must be made... between mere appreciation of the work the animal does, and the love of an animal because it is useful' argues du Boulay, writing about Greek country people and their relationship with animals (1974: 86). Du Boulay explains that animals are not loved for their 'sheer utility' but because they are 'useful' members of the rural household. And the rural household rarely includes 'non-working' members. Thus animals, by means of their inclusion in or membership of the household, enter a relationship of 'mutual' or 'reciprocal obligation', according to which, like any other household member, they are expected to contribute to its welfare, being entitled in turn to the necessary care needed for their maintenance (ibid.: 86–89).

Du Boulay (1974), by recognising the inclusion of animals in the rural household, and their 'lowest position' in it, sets the initial parameters for deciphering the expectations rural Greeks have of their animals and the meaning they attribute to the term 'usefulness'. Starting from this point, I shall explore the ways in which Vassilikiot farmers 'care' for their animals, the ways they punish or complain about them, the repetitive, simple, but exhausting tasks of their everyday interaction with them. It is my objective in this chapter to situate the relationship of Vassilikiots with their animals in the context of 'order' (*taxi*), which is applied by the farmers themselves and regulates any object, being, or activity in the environment of the farm, rendering concepts such as 'care', punishment and 'usefulness' meaningful.

The next section is an ethnographic presentation of the animals in question, that is, the animals 'kept' by the average household in Vassilikos. Reference is made to their basic husbandry and their locally defined usefulness. Then, in the subsequent section, I proceed to examine Vassilikiots' engagement with systematic animal husbandry and the specifics of work invested in the maintenance of 'flocks of animals' (sheep). Following this, I clarify, by means of further ethnographic examples, the meaning of 'order' (*taxi*) and 'care' (*frontidha*), two local concepts regulating the relationship of farmers to their animals. 'Order', in particular, is a central indigenous concept pertaining to the human-animal relationship in Vassilikos, since it embraces and directs the content of several other sets of meaning examined in this chapter.

Caring for animals on the farm

Vassilikiots apply the term 'animals' (*zoa*) to 'their' animals on 'their' farm. This does not mean that 'wild', undomesticated animals are not entitled to the term 'animal', but Vassilikiots are mainly concerned with their own animals, 'their' farm animals. In a similar way, while all animals on the farm are called 'animals', the term is more often applied to sheep and goats. For example, in the context of any given conversation, a farmer will refer to chickens and dogs as 'chickens' and 'dogs' – that is with their generic names – and to sheep and goats (and occasionally cattle) with either their generic name or simply the term 'animals'. Here, the generalising term 'animals', does not indicate negligence or disregard for the animals in question. On the contrary, it suggests an implicit recognition of their contribution to the wellbeing of the farm. In this respect sheep and goats are 'animals' proper.

Sheep and goats are typical examples of what the local farmers consider to be useful farm animals. They are common, present on almost every farm, and are an indispensable unit of animal stock held by the average household in the village.

In the past, four goats and four sheep were usually kept by every family. Some families even had a cow for milk. Nowadays, it's more or less the same. We all keep, at least, a couple of goats. Even an old man like myself.

As the words of this elderly Vassilikiot suggest, the number of sheep or goats a household has depends upon the age of household members and the energy they can devote to caring for them. While most Vassilikiot families do not maintain 'flocks of animals', the great majority of them 'keep' (*kratoun*) a small number of female goats or sheep, which can be easily watched, grazing on the farmland adjacent to individual households. The adult animals are tethered to an iron stake with a five-metre-long rope. The stake is poked into a different piece of land everyday (cf. Greger 1985: 33). The animals graze this piece of land within the diameter of the five-metre rope. The young animals, kids or lambs, are left free to gambol and graze around their mothers. Before sunset the villagers 'gather' (*mazevoun*) the 'animals' back on to the farm.

Special state and EU (European Union) benefits encourage this kind of small-scale animal husbandry. The benefits are designed to subsidise animal husbandry and farmers consider them an important incentive for keeping a minimum of seven or eight sheep or goats on their farmland. This minimum number, which is often overstated (cf. Green & King 1996: 657), entitles them to the EU subsidies. In addition, the sale of kids or lambs at Easter provides some extra cash for the household's budget economy. Easter is when most of the young kids and lambs are killed, with the exception of 'those who are destined to stay alive' (*afta pou einai gia zoi*) to replace an older ewe or goat. It is also expected that one kid or lamb will be consumed in the household itself on the same religious occasion. Vassilikiots are proud to be in a position to consume the meat of animals they raised themselves. The quality of the meat is referred to as being exceptional and the household's self-sufficiency as a productive unit is directly or indirectly acknowledged by guests and family members alike.

All the Vassilikiots I know, unanimously, declare a preference for sheep over goats. Having read John Campbell's (1964) classic ethnography about the Sarakatsani before I went to the field, I couldn't help thinking about his remarks on the same topic, every time that a Vassilikiot man or woman compared sheep with goats (see also, Blok 1981: 428–30; Brandes 1980: 77–9; 1981: 221). For the Sarakatsani, sheep are 'God's animals'; they are 'docile, enduring, pure and intelligent' (Campbell 1964: 26). Goats, by contrast, are associated with a wide array of negative features: '[they] are unable to resist pain in silence, they are cunning and insatiate feeders... although Christ tamed these animals the Devil still remains in them' (Campbell 1964: 31). In Vassilikos, although goats are not disdained to the same degree, they are often blamed for their 'disobedience' and their 'untamed' character, while, at the same time, sheep are praised for their submissive and benevolent constitution. 'Sheep are more obedient' and 'more docile (*irema*) animals', Vassilikiots claim. Watching the kids playing and fighting with each other, they make comments like:

Look how unruly (*atakhta*) the kids are. They are strong and grow well, but they don't stand still even for a minute. The lamb is not as frisky (*zoiro*) as them. The sheep is a blessed animal! (*einai evlogimeno*).

But, at the same time, the farmers cannot hide their secret admiration for the kids' strength and good health. Goats, by being 'wilder' than sheep, are expected to be 'stronger', more 'resistant' to disease or harsh environmental conditions. This is how a local farmer puts it:

> Lambs are good animals but weak. They are very weak in comparison with goats. Last year I lost a few ewes because of an illness. I rarely lose goats to an illness.
>
> But you see I make cheese and I need to have all those animals. In order to make cheese you need sheep, otherwise the cheese is not good.

As this farmer suggests, the milk of sheep is better suited to cheese production and this is why Vassilikiot flocks are composed of sheep rather than goats. However, those Vassilikiots who are not seriously involved in animal husbandry prefer to keep some goats on their farmland, investing the minimum of care and worry in exchange for the meat, milk, or state benefits derived from them.

Dogs, like goats and sheep, are present on any Vassilikiot farm, signalling the appearance of strangers with their persistent barking. They are simply described as 'useful' animals by their owners. 'Dogs do work', Vassilikiots maintain and acknowledge the conventional role of dogs as guards. But an individual dog is primarily evaluated in terms of its contribution to hunting. 'It is a good dog, it hunts', the local farmers emphasise in order to justify the special attention and 'care' they devote to particular dogs (cf. Marvin 2000a: 112–114).[1] By contrast, dogs unsuccessful in hunting are relatively neglected: they are fed but spend endless hours tied up. However, the farmers often spare a few sympathetic words for these less fortunate animals. This is because a dog, more than any other animal on the farm, meets the expectations of a farmer in respect of the notion of 'order' (*taxi*). For the villagers, obedience and devotion are not merely stereotypical canine behaviour, but represent what one expects from every animal on the farm but very rarely gets.

There are not many cattle left in Vassilikos, although, as the farmers maintain, in the past there used to be a lot. 'Cows were for milk, but for ploughing as well (*iton gia gala, alla kai gia zevgari*)', they add. Nowadays, the old-fashioned, local variety of cows, which was used for both milking and ploughing, does not exist any more. It has been replaced by a hybrid, a cross between local cows and 'improved (*veltiomenes*) cows from abroad'; the latter are referred to by most Vassilikiots as 'those (*aftes*) which give (*kanoun*) more milk'. In local narratives of animal husbandry, the old-fashioned local variety of cows is portrayed as having greater 'endurance' (*antokhi*) or 'strength' (*dhynami*), being perfectly adapted to the requirements prescribed by the local understanding of a cow's 'usefulness' and the imperative of self-sufficiency.

The introduction of tractors in the 1950s and 1960s, however, rendered cattle unnecessary for ploughing. Now that milk production is more profitable, the 'foreign' varieties of cows have been introduced on the basis of their productivity. But the earlier, local variety of cow was not simply replaced, it was interbred with the new animals. Endogenous developments in European farming, as van der Ploeg and Long have argued, are part of a process that enhances heterogeneity and contains 'a specific balance of "internal" and "external" elements' (1994: 1–4). Vassilikiots strategically plan crossbreeding, since they believe that it adds to the 'strength' of their animal stock (cf. Marvin 1988: 88, 92). They usually prefer to crossbreed a newly acquired 'foreign' (*xeno*) animal with those already present, instead of replacing the older variety completely.

Poultry are ubiquitous among the livestock of the average Vassilikiot farm. Backyards and the nearby cultivated fields overflow with poultry of all kinds, but primarily chickens and turkeys. All these birds are left free to roam around the farmland and olive groves, preying upon worms, fallen olives, and any food they can find. In the evenings, they return to the farm to be sheltered and fed by the farmers. They crowd around their owners, who throw them some corn, wheat, or other kinds of grain as an additional supplement to their diet. Although Vassilikiots do not worry much about the safety of fully grown birds – predators like foxes do not exist on the island – they do devote a lot of time and concern to 'caring' for new-born chicks. Most hens lay their eggs unobserved in various hidden places on the farmland, but as soon as farmers notice their new-born chicks, they collect them and put them in cages along with their mother or a foster mother. There, the chicks are protected and fed well for a couple of weeks, until they are old enough to care for themselves successfully. During their first days of life chicks are considered to be at risk (*kindhynevoun*). They may become 'lost' (*borei na khathoun*), be killed by rats or die in a sudden storm.

Unlike the chickens, which are capable of 'hatching their own eggs', turkeys are believed to be 'stupid'. 'They are clumsy and often destroy their own eggs', Vassilikiots explain, 'they always go and lay their eggs away from the farm, where the eggs will definitely be damaged by rats'. This is why the farmer will follow the turkeys to their nests and return their eggs to the farm. 'The turkeys are so stupid, that they keep on returning to the same spot to lay another egg the next day' the villagers remark. When enough turkey eggs are collected, the farmers will 'set a nest' for the turkey to incubate, or entrust the turkey eggs to a hen who is presumed to be 'a better mother'. Turkey breeding is thought so problematic, that nowadays many farmers prefer to buy turkey chicks reared in an incubator.

Turkeys are raised for the sole purpose of being sold at Christmas, when they bring a significant profit to the household. In the late autumn months, Vassilikiot olive groves are filled with turkeys and their characteristic call can be heard everywhere. Unlike turkeys, chickens are valued for both their meat and eggs and are consumed throughout the year at celebrations or other special occasions, especially when the household members wish to honour a guest (cf. Friedl 1962: 31).

As I have stated before (chapter Four), Vassilikiots always take pride in consuming their own animals.

Geese and ducks are disliked by many farmers in Vassilikos. 'They eat like elephants and make the water dirty for other animals', the local farmers complain, 'they are unruly and don't let the chickens eat, unless you stop them!' Geese and ducks are only kept for their meat, but some Vassilikiots appear unenthusiastic about its taste. 'Their meat smells', they argue, 'this is because they spend all day in the mire and the mud'. In spite of all these negative attributes there are a few farmers who keep a few of these birds on their farms. They merely 'keep' them (*ta kratoun*) either because they feel that a farm must have all kinds of animals on it – an aesthetic variant of the self-sufficiency ideal – or because, they argue, they have got used to raising them 'all those years!'

Like poultry, rabbits are numerous on some Vassilikiot farms. A few are left free to roam around in a semi-wild condition. But most of them are reared in cages and fed by the farmers with special care. 'Rabbits are weak animals', Vassilikiots maintain, 'those which are free eat any kind of food and often die of disease'. Rabbits are raised for meat, which is consumed throughout the year at celebrations or when guests are present. Vassilikiots boast about their '*stifadho*', a particular recipe for cooking rabbits or hares. During the summer, rabbits and chickens raised on the local farms are cooked in the local *tavernas* or restaurants. In most cases, the same households that own tourist enterprises are in a position to raise chickens and rabbits on their farmland. In this sense, tourism and farming appear as complementary manifestations of an economy centred around the household (cf. Theodossopoulos 1997b: 253–4; 1999: 3–4; see also, Chapter Four).

Pigs, like turkeys and rabbits, are reared solely for their meat. Like turkeys, the time of their death is well specified in advance. As soon as a young piglet is acquired, it is prescribed to die on a particular occasion. The rest of the pig's life will be a period of continuous fattening. If pigs have a particular privilege over other animals on the farm, it is that they are expected, and indeed encouraged, to 'get fat'. But unlike turkeys and rabbits, which fulfil a more integral and central role in a farm's yearly cycle, pigs are never found in great numbers and rarely multiply on the farm. The farmers avoid long-term pig breeding because, as they explain 'those animals smell' (*afta ta zoa myrizoun*). Since most Vassilikiots maintain room renting facilities or other tourist enterprises on their farmland, pig husbandry is obviously not a very wise economic strategy. One Vassilikiot woman explains:

> Raising pigs can be profitable. They bear a lot of young, up to fifteen sometimes, and you can rear them and sell them for fifteen thousand *drachmas* each. But they smell... They smell a lot! Some years ago I had a few. But then, because of tourism... If you rent rooms and have tourists close to your farm, you can't have many pigs.

To end this detailed account, I will briefly mention what Vassilikiots have to say about horses and donkeys, typical examples of 'useful' animals, which are deprived of their traditionally defined 'use'. In the past, horses and donkeys were used in transport and, like cattle, in ploughing. The older Vassilikiots vividly recollect transporting locally produced goods to the town with carts, a five-hour-long journey on a dusty dirt road. They also reflect on their labour and 'sweat' in the fields while 'doing *monaletro*', which means ploughing with one animal, a horse or a donkey. With the widespread introduction of tractors thirty years ago, horses in Vassilikos became redundant and their numbers declined. Stripped of their instrumental 'usefulness', however, they are still referred to in Vassilikos in tones of restrained nostalgia. Men are particularly delighted to talk about them, since, as they explained to me, 'riding horses and knowing about horses was the concern of men.'

Since there were not enough horses left, it was hard for me to investigate the relationship of men and horses in practice. I met, though, several young Vassilikiot men who advertised their experience or knowledge 'about horses': '... we grew up with carts, with horses. This is how we know how to saddle a horse and many other things that a horseman (*alogas*) knows...' As the following short narratives demonstrate, tourism has recently provided new economic incentives for some people to keep horses and donkeys on their farms. This is related to the passion of tourists for riding and the commercial success of 'folkloric' images associated with 'authentic peasant' lifestyles. This is how Vassilikiots put it:

I have this old mare, as you can see. She is old and unable to breed. She is of no use any more (*dhen khrisimevei se tipota pia*) and her food is costly...

In the summer I gave her to those people who organised a riding school for the tourists. They made some money but they gave us nothing. They promised me a new saddle, but...

You remember old Michalis's donkey, don't you... That miserable old thing that spent the whole winter in the olive grove opposite your house [see the photograph below]. I bought him for very little money, I cleaned him up and fed him. He became young again! You wouldn't have recognised him.

Then, I organised this 'Greek night' at my bar. A lot of people came, and a lot of tourists. Then, someone dressed in *tsolias* [traditional male costume with the characteristic white shirt] rode the donkey. He rode all the way through the village. The people were cheering and the female tourists were fascinated!

Figure 4 Michalis's old donkey before engaging in a new career in tourism

Keeping 'flocks of animals'

Some small 'flocks' (*kopadhia*) of 'animals' (*zoa*), sheep and goats, have always been kept on the land at Vassilikos in the recent past. The average size was between sixty and one hundred animals. In the last thirty years, few people were willing to keep flocks (cf. Just 2000: 53), and shepherding was considered to be among the least rewarding occupations one could have. In the 1990s, however, new economic incentives, like state and EU benefits and the growing tourist demand for locally produced cheese, made animal husbandry attractive to a new generation of Vassilikiots. This is how a local shepherd reflects on the state of the affairs:

> There is plenty of pasture in Vassilikos, but in recent years the flocks were few. There was a big flock in Xirokastelo, one in Potamia, one or two on the plains of Vassilikos. Now, the flocks are increasing again. A few people who had twenty animals or so enlarged the size of their flocks.

As is suggested here, some Vassilikiots are increasing the size of their animal herds. Since most local families already keep a few goats and sheep on their farmland, the transition to the stage where the animals form a 'flock' (*kopadhi*) is a gradual one. At some point certain farmers might decide to kill fewer lambs for a few successive seasons, thus increasing the size of their flock.

Vassilikiots owning flocks of animals are locally referred to as people who 'have [own] flocks' (*pou ekhoun kopadhia*), and only rarely as 'shepherds' (*voskoi*). Most of them proudly claim that they 'know' about 'animals', and have some on their farms. A farmer's decision to form a flock and devote attention, time and energy to 'caring' for it does not sever the farmer's relationship with other kinds of farming activities, such as the keeping of other farm animals or participation in the olive harvest. The household-oriented character of the local economy makes this feasible, since some members of the family can take 'care' of additional 'farming' responsibilities, while others, primarily men, stay out in the fields shepherding the flock. If the farmers realise that their venture into sheep and goat herding is economically unprofitable and 'they cannot make it' (*dhe vgainoun*), they will simply sell most of their animals and resort to other forms of farming, or even tourism.

Until thirty or forty years ago, that is when most Vassilikiots were landless, most flocks of animals were owned by the landlords. The labourers (*kopiastes*) who were in charge of the flocks, were entitled to some proportion of the animal products: cheese, milk, or cash from the lambs killed at Easter. The exact ratio of the labourer's share was defined by *sempremata*, the set of rules defining the economic relationship between landlords and tenant farmers living and working on a landlord's land.[2] *Sempremata* relating to animal husbandry, like those on agriculture [see Chapters Three and Five], were particularly disadvantageous to the labourers, allowing few opportunities for them to accumulate wealth. In the case of flocks, for example, the labourers were responsible for a number of sheep, or 'heads' of animals (*kefalia*). The labourers had to manage the number of lambs killed at Easter, so as to maintain or 'keep' the original number of 'heads' entrusted to them by the landlord. Any loss of animals, due to accident or illness, was charged to the labourers and it was blamed on the labourers' lack of 'care' or 'concern' for the animals in question. As Vassilikiots vividly recollect 'any time an animal was lost the landlord used to say: it was your fault, you didn't care for the animals well enough!' A sixty-five-years old farmer elaborates on the same topic:

> I always 'kept' animals. Thirty of the master's and not one of my own. When I asked the master if I could keep a ewe-lamb (*miliora*), he said: 'Not even a cockerel of your own will you have as long as you live on my land.'
> In the 1970s I got land of my own. Now, I have land and animals, but I can't do much. In the past I could do a lot, but I had nothing...

Some landlords, those portrayed by the local people as 'the good masters' (*oi kaloi afentes*), used to 'allow' the labourers to 'keep' some animals of their own in addition to the number of 'heads' entrusted to them in the first place.[3] As soon as the landless tenants of Vassilikos started acquiring their own land, they became more independent, and succeeded in negotiating better terms or better *sempre-*

mata in their arrangements with the landlords. Having animals as *misaka* (: half ownership) was one such arrangement. But nowadays, although all people who maintain 'flocks' depend – to a greater or a lesser degree – upon a landlord's land for the grazing of their flocks, the animals comprising the flocks are their 'own' property (*dhika tous*). A forty-year old farmer points out: 'Why should I spend so much effort to keep those animals, if I have to keep them as *misaka*! Only somebody stupid would do so much work for the sake of the landlord nowadays!'

As I have already mentioned in the previous section, goats are present on almost every farm in Vassilikos, but are rarely numerous enough to form flocks. With the exception of one herd of goats in Xirokastelo – the mountainous region adjacent to Vassilikos proper – the Vassilikiot 'flocks of animals' are primarily flocks of sheep. The rationale of this preference for sheep – apart from the more general Vassilikiot admiration for the sheep's constitution discussed in the previous section – is that the milk of sheep is indispensable for the production of good quality cheese. As a local 'flock' owner explains: 'The sheep makes less milk than the goat. But the milk of the goat is not as good... When you make cheese with a lot of goat's milk the cheese smells.' This is why only a small amount of goat's milk is mixed with that of the sheep in cheese production. The local variety of cheese, *ladhotyri* (literally, cheese in oil), is a traditional Zakynthian product, and it is popular, not only among the local population, but among tourists, both foreign and Greeks from the mainland. Vassilikiots owning 'flocks' are heavily engaged in *ladhotyri* production and the profit from it is a serious incentive for maintaining the flocks.

Vassilikiots recognise a special local breed of sheep, called 'the Zakynthian sheep' or simply the '*dopia* (local) sheep'.[4] *Dopia* sheep are larger than those from mainland Greece, have longer necks and curved noses, and are locally considered to be more beautiful. However, *dopia* sheep are frequently interbred with sheep from the mainland, or even with foreign varieties, since their milk production is not considered to be sufficient. 'The *dopies* ewes don't give so much milk; the crossbreed ones (*bastardhemenes*) are the best ones' the local flock owners maintain. *Bastardhemenes* means 'bastard', or of 'mixed breed'. Vassilikiots, as I have already mentioned in the previous section, prefer to crossbreed newly introduced breeds with the animals from their flock, instead of merely replacing the old breed with a new one. In this way they feel that they can better control the attributes of particular breeds: milk production, 'strength' or even appearance (cf. van der Ploeg & Long 1994: 1–4). The success of Vassilikiot inter-breeding strategies depends on careful observation and a vast knowledge of the attributes of most individual animals in the vicinity. All Vassilikiots 'who own flocks' are in a position to recognise the 'animals' of neighbouring flocks. Some of them proudly maintain: 'We know all the sheep in Vassilikos, their history, who their mother was, to whom they belong.'

An elaborate variety of names is used by Vassilikiots to refer to their 'animals', sheep and goats – terms that share an affinity with 'cattle descriptives' identified by ethnographers among African pastoralists (see for example, Galaty 1989:

219–25; Evans-Pritchard 1940: 41–5) or names given to fighting bulls and cows in Spain (Marvin 1988: 94, 97). Locally standardised names denote particular animal characteristics, like their colouring and other physical features and facilitate the identification of particular animals in a given flock of sheep or goats. They further facilitate conversations about animals between fellow-villagers, since they directly portray the appearance of the animals in question. To my knowledge, similar sets of standardised names of sheep and goats exist in most provinces of rural Greece. Individual animal names, like *Giosa, Liara and Bartsa*, are widespread and commonly used in many places, but the majority of names 'for goats' (*gia ta gidhia*) or 'for sheep' (*gia ta provata*) represent innovative expressions of local culture and are influenced – at least in the Zakynthian case – by regional dialects. Table 1, lists a catalogue of these names as I recorded them in Vassilikos.

TABLE 1

Names 'for sheep' (*gia provata*) refer to either gender (*zygouri/kriari*: a ram or *provata*: a ewe) and age (*arni*: a lamb or *miliora*: a ewe-lamb) of the animal in question or other physical characteristics, like small ears (*Tsipa*) or possession of horns (*Kourouta*). But the majority of the names 'for sheep' refer to the colours of female sheep (*provates*: ewes). Here is a short catalogue:

Liara – a white ewe with black spots.
Belitsa – a completely white ewe.
Mourtzina – a white ewe with white and black spots on her face.
Gardelha – a white ewe with various colour patterns on her face.
Mavromata – a white ewe a with black-coloured-spots around her eyes.
Katsena – a white ewe with brown-coloured patterns on her face.
Lagia – a completely black ewe.

Similarly, some of the names 'for goats' (*gia ta gidhia*) refer to physical features like lack of horns (*Souta*), small ears (*Tsipa*) or horns that turn backwards or upwards (*Pisokera* and *Orthokera*), while others refer to the coloured patterns of she-goats and are comparable – often identical – with the ones applied 'to ewes':

Liara – a white goat with black spots.
Layia – a completely black goat.
Mora – another name for black goats.
Bartsa – a white and grey goat, which is 'rather white' (*asprouliara*) on the front or
     middle part of her body and black at the back.
Giosa – a white goat with grey markings on her body.
Koukia – a cinnamon-coloured goat.
Khiona – a completely white goat.
Kokkino – a goat of a reddish colour (*kokkinopi*).
Boutsika – a goat with greyish colour patterns on her face (*psari sto prosopo*).
Rousa – a somewhat reddish or yellow goat, and rather large.

Vassilikiot flock owners identify two major factors that determine the economic viability and success of 'flock' husbandry in Vassilikos. The first is related to the availability of household members to contribute to the 'care' and labour related to the flock. Young and older women provide valuable help in milking and cheese production, while boys and older men often replace young men in sheep herding, in case of illness or in the absence of the main shepherd. Vassilikiots who are in their late twenties or thirties appear confident about their engagement with animal husbandry. They are optimistic about the future of their flocks and gradually increase their size year after year. They all have young partners and active parents, who offer valuable help to them. But the older flock owners in Vassilikos, despite their clearly expressed concern for their 'animals' cannot hide their exhaustion. They have to rely on the assistance of their sons or daughters who are not always available or willing to help. In all cases that I recorded[5], however, the existence and welfare of the 'flocks of animals' in Vassilikos relies upon the corporate, collective orientation of the household economy: the willingness of the household members to co-operate, realise 'self-sufficiency' and maximise the household's resources.

Access to land for pasture is the second major prerequisite for the maintenance of flocks of animals. Since none of the flock owners in Vassilikos has enough land to satisfy his animals' appetite all the year round, all of them have to secure pasture on other people's land, through generalised networks of obligation (in case of a landlord's land) or reciprocity (in case of a fellow farmer's land). Access to those pastures could be on a long-term basis, as in the case of a tenancy agreement (*sempria*) traditionally 'given' by a landlord to individuals from particular families, or on a temporary basis, as with more ephemeral agreements between two neighbours. The latter kind of arrangement clearly entails a reciprocal element. 'I prune Dionysis's olive trees and my animals eat the leaves of the cut branches (*tsimes*)', a young flock owner explains and adds, 'in this way we both have some benefit... I feed my sheep and he has his trees well pruned...' Vassilikiot flock owners carefully respect agreements relating to animal pasture. In cases of trespass the consequent tension is, most of the time, short-lived and the local farmers, who are often neighbours, maintain their friendly or – as they say – 'good' relationships.

Notions of care and order

Conducting frequent informal tours of the farms of Vassilikos is one of the most informative and pleasurable activities for the visitor-cum-anthropologist, who thus enjoys a chance to get acquainted, not only with the physical surroundings, but also with the indigenous discourse on caring for and ordering the animate and inanimate environment. Ideas, beliefs and above all maintenance strategies are communicated to the visitor while the discussion is often fuelled by the physical presence of the animals, the view of well-tended vegetable gardens and carefully-

Figure 4 Mimis, one of the older flock owners, with his flock and sheep-fold

built animal shelters. Each and every one of these encounters triggers discussions that revolve primarily around the labour and care invested by the farmer. If, in turn, the visitor happens to have been on the farm before, then its owner will most definitely concentrate on recent changes that might involve newly acquired/born animals, projects currently being undertaken, and, above all, the planning needed in order to materialise the dream of the perfect farm. A very similar scenario unfolds when the visitor is a knowledgeable neighbour and a fellow farmer. In this case however, the discussion is most likely to concentrate on instrumental aspects of animal care and cultivation since Vassilikiots are usually eager to share their knowledge and skills with others. So much so, that any new ideas and techniques related to animal husbandry and cultivation are effectively disseminated around the village with unique speed and down to the last detail.

Comments on the animals' behaviour, critical information about diseases and inspired changes or innovations in the construction of animal shelters are the subjects that most frequently preoccupy the conversation of two fellow farmers. In the case of a neighbourly visit, the host will straightforwardly express his or her pride and satisfaction in the wellbeing of the farm, since in Vassilikos the orderly arrangement of animals and structures is regarded as an achievement to be solely credited to the farmer and the farmer's family. The visiting farmer, on the other hand, acknowledges the hard work invested in the place by praising the host and expressing his or her admiration, always from the standpoint of the connoisseur who is in a position to recognise and appreciate another farmer's accomplishments. The effective organisation of the farm and the projects scheduled to be undertaken in the near or not so near future are usually the main concerns of such discussions. Problems relating to the practical requirements of running a farm and the 'care' of particular animals are also an integral part of those peripatetic conversations, which can prove to be particularly informative and instrumental for farmers and the anthropologist alike. For the former acquire useful suggestions on frequently encountered problems and the latter gains abundant ethnographic insight.

The farmers' narratives go back in time and are characterised by an almost 'biographical' element: past experiences are reflected in the present and projected in the hopes for the future. In the course of a farmer's narration one can visualise the state of the farmland when it was bought from the landlord. Through the detailed recounting of how, and in what manner, each aspect of the farm was developed, the visitor is given a chance to evaluate and admire the changes that have taken place over the years. If not a farmer, the visitor can only imagine how much effort was required for the achievement of the present day ordered state of 'things'. This notion of 'order' and hard labour, central in the farmers' discourse about the transition from the past to the present, is also intimately linked to the 'days to come'. By pointing to empty plots of land and describing new shelters for animals not yet born, or by explaining how this or some other vegetable garden will be better 'fenced and watered', the farmers amply demonstrate a sense of continuity that makes the farm seem like a living, breathing organism with its own distinct history and future.

Indeed, order is the organising concept that connects the past and the present with plans about the future: ideas for a better organisation, new cultivation to be introduced and more animals that have to be 'cared for'. Safeguarding this order, is a constant responsibility for the Vassilikiot farmer. A characteristic preoccupation of this kind concerns the removal of undesirable vegetation, a never-ending struggle since nature constantly regenerates itself [see Chapter Four]. Similarly, cleaning and repairing the animal shelters, maintaining and improving the fences of vegetable gardens, repairing all material constructions subject to damage by either animal activity or the weather are all part of the repetitive duty of preserving the order of the farm. Time, weather and the persistence of weeds might cause constant trouble for the farmers, but are actually easier to handle than the chaos some of the animal members of the farm can cause.

Vassilikiots consider domestic animals to be prone to disorder, especially when left unattended. Not surprisingly, they treat them in a manner comparable to that in which adults – in this particular cultural context – deal with their young children. Animals are thought – much like children – to be unable to survive on their own. The farmers' constant 'caring' presence, intervention and control are hence deemed to be of decisive importance to the animals' well-being and the prosperity of the farm as a whole. 'Teaching the animals their place', to use an expression so frequently used by the Vassilikiots, is locally considered an extremely important task. Animals are frequently punished and rewarded for violating or complying with the farm's order respectively. As the following ethnographic examples illustrate, the Vassilikiots believe strongly in the ability of domestic animals to learn and their own well-tested and therefore infallible educational strategies.

A great part of my time in Vassilikos was devoted to animal care and helping the local farmers in their daily routine. Gathering the household's sheep and goats from the various parts of the farm where they were roaming was one of those afternoon tasks that helped me realise the labour involved in animal husbandry. I found this job particularly tiring – wondering all the while at the way people decades older than me did it so skilfully – not simply because it was repetitive, but primarily because it involved a fair amount of walking across rough ground and considerable effort pulling the animals along by their tethers. Some animals, especially younger ones, tend to be disobedient and their unruliness adds a certain amount of extra difficulty to the task. Furthermore, as the farmers themselves explained to me, individual animals can be stubborn. They might refuse to get in the pen or, even worse, to remain in their appointed 'place' (*tin thesi tous*) within it. Most of the farmers are adamant in their expectation that the animals should 'learn' (*na mathoun*) their 'right place' in the pen, employing punishment as an effective teaching method.

Punishment, then, consists mainly of beating and shouting at the animals, as well as trying to reason with them. 'Why don't you stay in your place?' and 'How many times do I have to teach you your right place!' are some of the most frequent remarks directed at animals who refuse to respect 'the order of things'. Younger animals are expected to disobey more frequently and are thus punished more often. But even those, the farmers maintain, 'learn in time' to respect their defined or 'right' place on the farm. It must be noticed, however, that not all species of animals are – or are expected to be – equally disobedient. In this respect, goats are the prime suspects for misbehaviour [see my earlier commentary in this chapter]. They tend to 'disobey' the farmer more often than the sheep and are consequently punished with greater frequency. While beating their goats, the farmers tend to compare their boisterous behaviour with the blessed submissiveness of the sheep: 'Look how the ewe knows its place. Goats are not like that. Neither is the ewe lamb, but it will learn in time'.

Orphan kids and lambs in Vassilikos are suckled by foster mothers, nanny-goats and ewes. Some of them accept the foster kid or lamb and care for it, but others, especially those which already care for their own young, strongly resist suckling orphans. The farmers recognise that it is 'every mother's instinct to feed her own child', but at the same time they maintain that all animals on the farm 'must' receive 'proper care'. This is why, nanny-goats and ewes which deny the teat to orphan kids and lambs are punished for their resistance, while those which agree to feed and foster orphans are praised by the farmers for being 'good mothers' and 'good animals'. The latter comply fully with the farmers' demand for 'order' and 'self-sufficiency', since they succeed in providing the maximum 'care' with the means already available on the farm.

Despite the popularity of punishment as a controlling method, I was never a witness – like du Boulay – of 'deliberate cruelty to animals' (1974: 89). Beating and shouting at them always takes place in the context of safeguarding the order of the farm. The most common causes for punishment are either intrusion of the animals into forbidden places, like vegetable gardens and barns, or physically harming another animal or eating its food. The Vassilikiots express particular distress when they 'have to' (*ekhoun*) penalise their animals and discipline is always accompanied by admonishing the defiant creatures. The villagers talk and scold the animals as if they were children, explaining to them their errors and transgressions: 'I'm rearing you! Why don't you listen? Why don't you learn your place?,' they often shout at those animals who refuse to quietly assume their predetermined positions in the shelters or cages. What Vassilikiots find especially frustrating, as they confided to me, is the inability of some individual animals to acknowledge that confinement in the shelter is aimed primarily at protecting them from weather conditions and wild predators. In situations like this, they often express their disappointment by appealing to the unruly animal with statements loaded with emotions: 'I am rearing you... I am rearing you' (*ego sas anastaino*), they repeatedly exclaim.

In conversations about animal care, the assumption that all animals in Vassilikos have 'somewhere' to sleep is implicit. Those animals that are regarded as more vulnerable to disease – cows or rabbits, for example – enjoy more carefully designed shelters, while other animals that are thought to be more resilient are accommodated in more elementary and temporary structures. In both cases, however, sheltering all animals adequately is an essential constituent of 'order' on the farm. The sight of farm animals wandering freely around the farmstead at night is considered by the farmers the epitome of disorder.

Smaller animals, chickens or rabbits, that suffer from disease or an accident are usually killed by Vassilikiots themselves who see this as a form of euthanasia that relieves the unfortunate creatures from undesirable and possibly unbearable pain. Larger animals receive some basic form of veterinary care, which in serious illnesses consists of vitamins and antibiotics in the form of injections or capsules mixed in their food. Vassilikiots rarely resort to veterinarians since, as they claim, 'they know about' (*xeroun*) or 'can tell' (*xekhorizoun*) the most frequent and common diseases their animals are bound to suffer from. The most serious of those are treated with medication obtained from the town, but less critical conditions are often dealt with traditional remedies handed down from the farmers' 'forefathers'. This is how I learned, for instance, that 'camomile and oil make the ewe's stomach move again', or that 'ash from reeds mixed with water makes a horse's wound heal.'

Much like punishment, killing an animal is a crucial moment in the farmer's relationship with the non-human members of the household. An animal's death is always considered in relation to its contribution to the economy and wellbeing of the farm and it occurs when the 'order' of the farm dictates that an end should be put to an otherwise long established process of 'care'. As I have already stated, the farmers believe that the animals cannot exist without the security, 'care' and 'order' provided by themselves in the farm environment. In this context, the death of an animal is seen as a reciprocal bequest for the 'care' it received in the past. This kind of reciprocity has also been recognised by du Boulay who described the relationship between animals and their owners as a 'reciprocal' or 'mutual' one (1974: 86). To take this a step further, I would argue that both the 'care for' and the 'death of' an animal are phases or expressions of the 'order' of the farm.

When killing their animals the farmers in Vassilikos express distress and sorrow (*stenokhoria*). They often try to rationalise their emotions with jokes, or better, to avoid them by hiring another villager to 'do the slaughtering' of their own animals. Large animals are slaughtered by men (cf. Handman 1987: 152; Kenna 1992b: 167), but a few men actually specialise in this task and are locally respected for 'they know how to kill an animal quickly' and 'painlessly'. Small animals are almost invariably killed by the farmers themselves and while both men and women know 'how to kill' such animals, the plucking is done mainly by women since it involves kitchen utensils and is therefore deemed to be a female job. Chicken and rabbits, the kind of small animals that the farmers themselves would kill, are slaughtered on the spot and on any occasion that their meat is required, be this a planned celebration or an unexpected visit from a friend.

An animal's death is, without fail, dictated by some form of practical necessity articulated by the farmers in the context of a discourse that underlines the mutual interdependence of the constituent parts of a farm. 'If you have animals, you have to kill them as well... There is no other way... How else are you going to get the food to feed the rest of the animals?' the Vassilikiots reason. Managed by a household-centred economy and along the lines of self-sufficiency, the farm is locally perceived as a closed system with a specific hierarchy, at the top of which are the farmers themselves. This position gives the human protagonists the obligation of caring for the animals, as well as the sad duty to terminate a caring relationship when particular circumstances dictate that they should do so. As a Vassilikiot once told me: 'This chicken will die in eight months, this tree in a thousand years, there is a time for everything to die.' The longevity of domestic animals is then part of the articulation of 'care' and 'order', the two principles that govern and safeguard the wellbeing of the farm. According to those principles, the death of an animal only occurs after it has been well looked after, or to use the Vassilikiot expression, 'after it had a good life.'

Figure 6 Lefteris in the struggle of animal care

Conclusion

'Animals are not loved for themselves as members of the animal kingdom with their own beauty and peculiarity, but nor are they thought of in crude terms which involve only total exploitation of their productivity' (du Boulay 1974: 86).

At the beginning of this book I have described how groups of environmentalists have penetrated the Vassilikiot political scene, in a twenty-year-long effort to protect rare species of animal and establish a National Marine Park in the locality. The environmentalists, who are referred to by Vassilikiots by the generalising term 'the ecologists', present themselves as individuals who 'feel' (*noiazontai*) for 'nature' and its living constituent parts, the animals. They claim – to use the words of du Boulay quoted above (1974: 86) – to love animals as 'members of the animal kingdom with their own beauty and peculiarity' and accuse Vassilikiot farmers of thinking of animals 'in crude terms which involve only total exploitation of their productivity'. It is not surprising, then, that the environmentalists who campaign for the protection of animal species in Zakynthos, but who are ignorant of the true nature of the relationship that rural Greeks have with their animals, have made themselves particularly unpopular in Vassilikos and the surrounding communities.

The environmentalist's difficulty in appreciating the caring potential in the relationship of farmers with their animals is partly related to the reluctance of the farmers themselves to explicitly articulate the extent of their labour invested on a daily basis in animal welfare. Animal husbandry is often only one of many 'concurrent strategies of subsistence' (Hart 1992: 68) and repetitive cycles of work devoted to it are often understated by the farmers themselves who are busy managing more than one farming-related activity at a time. Rural Greeks, in particular, appear reluctant to use terms such as 'love' to describe their relationship with domestic animals, thus occasionally, leading outsiders, such as urban dwellers, or even anthropologists, astray with their projected emphasis on the practical dimension of animal care. Ernestine Friedl (1962), for example, in her classic ethnography about a Greek rural community, Vasilika, seems to underestimate the care of the local people for their animals. 'The villagers do not give their animals individual names,' she argues, 'they take no particular care to keep them physically comfortable' (1962: 30). Friedl refers to the 'beating' and 'kicking' of animals at work and the children's 'teasing' of them. She recognises that dogs and other animals 'are not considered pets,' but she describes the local peoples' attitude towards them as being 'completely utilitarian' (ibid.: 32).

Unlike Friedl, Campbell, in his well-known study of the Sarakatsani shepherds, acknowledges the importance of the human-animal relationship, which in his view 'must be seen not only in terms of utilitarian satisfaction or social function' (1964: 34). For the Sarakatsani 'shepherding has intrinsic value'; their conception of time and the organisation of their life revolves around the movements and needs of their flocks. The main concerns in the life of the Sarakatsani are 'sheep, children and honour', explains Campbell, and underlines the identification of the shepherds with their sheep, the latter being 'a prerequisite of prestige' (ibid.: 19, 30–1, 35). The Sarakatsan shepherds, like the Vassilikiot 'flock owners' discussed earlier in this chapter, are in a position to relate to the particular history and qualities of individual sheep and for this purpose they have developed 'an extensive descriptive vocabulary of sheep terms.' Sarakatsani 'care' for sick animals with 'compassion', Campbell observes finally: without being 'sentimental', 'an evident solidarity' exists between them and their animals (ibid.: 31).

The significance of sheep for the Sarakatsani is obviously related to their shepherding way of life. But most of Campbell's observations relating to the non-utilitarian, 'intrinsic' character of the relationship between animals and their owners, are in accordance with du Boulay's work (1974) and my own data in Vassilikos, both studies undertaken in farming communities. As I have already mentioned earlier in this chapter, du Boulay has recognised animals as the lowest members in the rural household, having, like human members, obligations and privileges of 'total loyalty and mutual support,' superimposed by a household-centred organisation of the village economy (1974: 16, 18, 86–89). She makes clear that animals 'occupy the lowest position' 'in the order of things' and in times of hardship are often expected to suffer more than, or at least as much as, the humans do, being the first to be sacrificed for the benefit of the household to which they are attached and bound by ties of 'reciprocal obligation' (ibid.: 86–89).

My ethnographic description of the relationship of the people of Vassilikos to 'their' domestic animals further supports the view that the relationship in question is understood as a 'reciprocal' one. The animals receive 'care' (*frontidha*) from their owners and the farmers in turn expect respect from the animals for the 'order' (*taxi*) of the farm, and that they should even sacrifice their own lives for its maintenance. The farmers clearly express in conversation the expectations they have of their animals and often talk to the animals themselves, despite their confident assertion that animals do not reason. They try to explain to them the 'order' of the everyday activities which directly concerns them, even the fact that their confinement under the terms of this 'order' is for their own benefit. The farmers of Vassilikos maintain that animals 'learn' (*mathainoun*), through repetition and punishment, their expected position in space and time, and my own observations suggest that most animals do 'learn' their place on the farm.

'Order' (*taxi*), as I have repeatedly illustrated in this chapter, is the prevalent central concept underlying most aspects of the human-animal relationship in Vassilikos. Punishment, 'care' (*frontidha*) and the termination of the process of 'care', the slaughter of an animal, are all different expressions of 'order' on the farm. Placed in this context, 'order' is directly related to the organisation of the household as an autonomous self-sufficient unit in opposition both to other households and the environment. 'Order' keeps household members, animals or humans, and the activities those members are involved in, well attuned to the self-interest of the household. Self-interest (*symferon*), as several other ethnographers have demonstrated (du Boulay 1974; Loizos 1975; Hirschon 1989), rather than being an expression of individualism, concerns the family or the household as a whole.

In the farm environment, 'order' is ideally maintained by the male head of the household, in a way that significantly resembles the responsibility for safeguarding family 'honour'. Similarly, in the domestic domain, 'order' is the primary concern of the *nikokyra*, 'the mistress of the house' or 'the female householder' (Dubisch 1986, Salamone & Stanton 1986, Loizos & Papataxiarchis 1991). In Vassilikos it is men, more often than women, who punish animals and take decisions concerning major issues related to animal husbandry and temporary or permanent buildings on the farmland. But women are usually responsible for the poultry, and participate in milking and various other everyday tasks on the farm. In their husband's absence or illness, women are capable of undertaking most jobs associated with animal 'care', even those related to the larger farm animals which would normally be expected to be a male concern (cf. Friedl 1967: 103–4; Handman 1987: 151–2; Hart 1992: 243–6; Galani-Moutafi 1993: 254). Consequently, the distinction between male and female spheres of responsibility on the farm represents the ideal of 'order', rather than its actual application, in a way that resembles the lack of 'isomorphism between gender roles and the domestic and public spheres' as argued by Dubisch (1986: 19) and Salamone & Stanton (1986: 98).

The farmers in Vassilikos are engaged in the repetitive, everyday tasks of 'caring' for their animals and 'keeping' their farms in 'order'. They feel they are themselves responsible for the wellbeing of their animals and their rearing, and openly express the belief that 'without them' and 'their struggle' everything would collapse into disorder. They design, define and safeguard 'order' on their farm and their right to do so is hardly ever questioned. It is well-supported by an elaborate religious cosmology which places human beings at the top of the hierarchy of living creatures. This religious theory about the creation and position of animals and human beings in the world will be discussed in detail in the following chapter. What I want to emphasise here is that the farmers in Vassilikos consciously present themselves as the indispensable, irreplaceable providers of 'care' and guardians of 'order' on their farms. They understand their role in relation to their farms and animals to be that of the ultimate 'caring principle'. This is why they express bewilderment, when they are accused by 'ecologists' or other urban dwellers, of being 'utilitarian' or 'exploitative' towards their animals.

Like the people they call 'ecologists', Vassilikiots firmly insist that they 'care about animals' (*noiazontai gia ta zoa*), 'their animals' (*ta zoa tous*). In their turn, they accuse the 'ecologists' of being unable to 'understand the struggle that [caring for] animals requires' (*dhen katalavainoun ton agona pou ekhoun ta zoa*). 'The ecologists don't know about animals', the Vassilikiot farmers explain, 'they talk about animals all the time, but they don't know about animals.'[6] 'We have animals and we know about animals' (*emeis ekhoume zoa kai xeroume apo zoa*), Vassilikiots argue and add: 'We live with animals and we know how to care about them.'[7]

NOTES

1 The great majority of Vassilikiot men are obsessed with hunting dogs and spend considerable amounts of time discussing them at any given opportunity. 'Dogs are useful animals', the local farmers maintain, 'they guard and hunt'. However, dogs that merely guard are provided with the minimum 'care' required for their subsistence; they are often fed on bread and water and are tied up for several days at a time. But dogs that excel in hunting are looked after conspicuously well. Their owners feed them well, worry about their health and overall condition, and most importantly, talk about or boast about (*kamaronoun*) them at every relevant opportunity.

2 As I have explained in Chapter Three, particular tenancy agreements or *sempries* were defined according to standardised patterns of *semberemata*.

3 '*Mas afinan na kratoume kai merika dhika mas zoa.*'

4 In Chapter Five, I referred to a local variety of olive trees in Vassilikos, which are similarly called '*dopies*-olive trees'. The term *dopies* is further employed by Vassilikiots to describe the 'local' breed of cows, as I have already described earlier in this chapter.

5 For detailed personal narratives of young and older Vassilikiot flock owners, see Theodossopoulos (1997b: 112–5).

6 '*olo milane gia zoa, alla dhen xeroun apo zoa.*'

7 '*Emeis zoume mazi tous kai xeroume pos na ta frontizoume.*'

7
CLASSIFYING THE WILD

Rarely do Vassilikiots' refer to wild animals and birds in contexts other than hunting. The threat of attack by small predators on the farm stock, or the occasional encounters with wild animals during their daily activities in the fields, are the rare exceptions. They usually respond to questions about wild animals by evaluating the animal's qualities, such as the animal's capacity to do good or harm. They often start by examining the possibility of harm and finish by considering the possibility of benefit. Most of these discussions are bound to centre on the issue of whether the animal or bird is edible or not, and its role as game. In general, wild animals classified as game 'receive more attention from humans than other wild animals' (Marvin 2000b: 206), while the vast majority of Zakynthian men find any discussion about hunting particularly fascinating.[1]

In this chapter, however, I will be concerned with the Vassilikiots' relationship with wild animals as expressed in contexts other than hunting. During my fieldwork, I experienced great difficulty in collecting data of this kind. In general, Vassilikiots were reluctant, to say the least, to talk about wild animals *per se*. Unable to instigate such a discussion, I often had to wait for unsolicited remarks to be made, or simply hope for their reactions to the rare sight of wild animals. It was in these instances that I would grasp the opportunity to ask further questions. Vassilikiots' brief but colourful answers exposed confident evaluations of wild animals that clearly reflected the Vassilikiots' status as members of farming households responsible for the welfare of their farms and the safety of 'their own' domestic animals in particular. In their confrontation with those wild animals that pose a threat to the order of the farm, the Vassilikiots' authority to decide the fate of the animals was taken for granted. Their decisions, however, sometimes diverged from narrowly defined utilitarian considerations.

In the second part of the chapter I will attempt to shed some light on the Greek Orthodox religious cosmology, which directly or indirectly informs Vassilikiots' understanding of the usefulness of animals and their place in the world in relation to any given human protagonist. To avoid generalisations, and present, at the same time, a coherent religious discourse, I have decided to examine a particular religious text, which I consider to be the most representative authority source on Orthodox perceptions of the animal world. This is a text known as the *Hexaemeron* (the Six Day Period) or *Homilies on the Hexaemeron* and its author is St Basil the Great (*Megas Vassileios*), one of the most venerated holy fathers of the Orthodox patristic tradition. His homilies in the *Hexaemeron* were delivered with the intention of providing an interpretative theology. In the four homilies discussed in this chapter, the author's two aims are the explanation of animal and plant creation as defined in Genesis, and the development of relevant moral examples or metaphors which inform correct Christian conduct. Regardless of the author's intentions, however, homilies E, Z, H and Θ, comprise a coherent classificatory discourse. They reflect an analytical cosmological exegesis based on conceptual categories and hierarchies which organise relationships between living beings.

St Basil's contemporaries were as immensely influenced by his interpretation of Genesis in the *Hexaemeron* as subsequent theologians have been.[2] In an acknowledgement of his work, the Church service in his honour includes a hymn in which St Basil is venerated as 'one who studied and interpreted the nature of beings'. The authority of the author and the influence of his exegesis are the reasons for my using it as an illustration of religious teaching about non-human beings. I acknowledge, however, that St Basil's exegetical approach to the natural world, although indicative of patristic tradition, is merely one patristic source on creation. The relationship of human beings to the natural world is the subject of ongoing theological debate in Orthodox Christianity, which my strictly anthropological approach merely touches upon. My study of St Basil's taxonomic insights is strictly confined to those aspects of his discourse that directly illustrate a particular cultural approach to animal classification.

To facilitate my presentation of Vassilikiots' or St Basil's judgements on wild animals, fish or birds, I employ the expression 'criteria of usefulness'. The term refers to the tendency to evaluate non-human beings according to their perceived 'use', lack of use, or even 'harmfulness' for the farming community and humans. I have chosen the term 'usefulness' rather than 'utility' in order to emphasise the potential for practical 'use' that the term 'usefulness' contains. At the same time, I attempt to distinguish between the rigid sense of utilitarianism, implied by the term 'utility', and the more flexible and negotiable form of relationship enacted by my respondents in Vassilikos. The following sections will illustrate this further.

Talking about wild animals

Τσιπουδρέλο λιαναρή
που πατείς την γη και τρίζει
Tsipourdelo (Robin) so slightly built
you alight on the Earth, and it creaks.[3]

As I have already mentioned, collecting data on wild animals and birds in Vassi-likos is not an easy task. During my fieldwork, whenever I had some information of this sort, I used to share it with as many Vassilikiots as possible, hoping – as was usually the case – that each individual would add something new to my enquiry. This is why, once I had proudly rehearsed this couplet about the robin, I hastened to share it with Lefteris, my adoptive father in Vassilikos. Lefteris replied:

'Who told you that?'
    'Adas [: a nickname] did, at the coffee-house', I said eagerly.
    'I see that you are learning well. Do you know why "the Earth is creaking?" Because when the robin lands on the ground he moves his body up and down like a spring (*sousta*). This is why!'

We both laughed. Then Lefteris continued, adding more information of the kind I was eagerly pursuing:

'We sang the couplet when we were kids. We used to set traps made of reeds. Some-times we would catch fifteen of them or even more!'
    'Is it edible? I didn't know that', I remarked.
    'Yes, it is. If you can catch a lot of them. Nowadays nobody cares. It is such a small bird, and does no harm (*dhen kanei kako*). It just needs moisture and worms. So it is easily deceived by the worm attached to the trap. Other times we used to dig up the ground a little so as to entice the robins into the trap'.
    'Are there plenty of them in Vassilikos? I haven't noticed any'.
    There are. Tsipourdeloi [robins] are gone during the summer (*einai fevgatoi*). They go to Bulgaria, Romania..., not like the sparrows who are locals (*dopioi*)'.[4]

Fifteen years ago the people of Vassilikos were made aware of a scheme for sea tur-tle conservation, something organised by outsiders. Bewilderment was their initial reaction. 'What use is the turtle?,' the local people wondered.[5/6] This is a question they still pose, despite persistent messages from the mass media and elsewhere[7] stressing the ecological significance and uniqueness of this particular animal species. The local people's attitudes towards the sea turtles, before the appearance of the environmental groups locally referred to as 'ecologists', were characterised by a passive and mute indifference. A fifty-years-old Vassilikiot remembers:

The turtles were never disturbed by us. When I was fourteen-years-old I used to pass through Gerakas [a beach where the turtles lay their eggs] leading animals, goats or even cattle, but nothing bad (*kako*) ever happened to the turtles. There were many of them at that time. Sometimes the waves might wash ashore a dead one that was giving off a stench.

Other Vassilikiots stress that people on Zakynthos do not eat the turtles and insist that 'the meat of the turtle is hard and tasteless'. Unlike other species of turtle, which are edible, the Loggerhead turtles that lay their eggs in Zakynthos are unappealing to the locals. This is why the farmers of Vassilikos, people who work the land and do not share a particularly strong affinity to marine animal life, appeared – before the recent introduction of turtle conservation on their land – totally uninterested in this rare reptilian species. 'The turtles were of no significance for the life of the people', a more reflective informant explained to me. 'They didn't do any harm, they were slightly useful, one might say... their eggs were food for the dogs.'[8]

The local view on another rare marine species, the monk seal, which is similarly a target of ecological conservation, is more clearly expressed.[9] 'This animal does harm' (*kanei kako*) they declare with indignation, lifting up their damaged nets for everyone to see. Large holes in the nets are the proof of the damage (*zimia*) caused by seals. During my fieldwork, I recorded two incidents of Zakynthian fishermen attempting to shoot seals despite the strict prohibitions imposed by the conservation regulations of the Marine Park. It may be that recent attempts to shoot the seals represented a form of challenge to the environmentalists on the island. Despite this possibility, however, most villagers in Vassilikos express their resentment against this particular marine mammal: 'Seals are and always were (in the past – prior to the ecologists' arrival) undesirable (*anepithymites*).'[10]

Talking about birds of prey, the people of Vassilikos emphasise the 'harm' (*to kako*) that these birds do to chickens and small animals on the farm. They also differentiate between edible birds of prey and inedible ones. The peregrine (*Petritis*), the sparrowhawk (*Xefteri*), and the goshawk (*Barmpouni*), are all edible birds of prey.[11] They mostly feed on birds that they kill while flying. This is what Vassilikiots call 'clean food' (*kathari trofi*) illustrating with examples: 'The peregrine (*Petritis*) is very, very proud. He only eats what he can catch in the air. If his prey were to drop to the ground, he would not fly down to pick it up'. They also refer to some of the criteria rendering a bird of prey edible:

You consider whether the bird's meat stinks, or if the meat is tasty, or if the bird is big enough. But what is more important is to see what the bird eats. Does the bird eat mice or carcasses or garbage? This is not clean food (*kathari trofi*).

Other birds of prey, like the lesser kestrel (*Kirkinezi*) and the black kite (*Loukaina*) are not considered edible for the reasons stated above.[12/13]

Vassilikiots are capable of naming nearly all the birds living permanently on or migrating to their land; they even recognise those birds which fly over their island for a short period on their migration route. They use local names, characteristic of the Zakynthian dialect, or names common throughout Greece. Although women do not hunt, they are equally capable of recognising and naming birds, especially women over thirty-five-years of age. They have a close practical experience with hunted birds, since plucking and cooking is locally considered to be 'a woman's job'. While preparing the birds for cooking they often find small animals or nuts in the birds' intestines and gain additional knowledge of the birds' diet. Pairing this further task with their observations of what birds eat in the natural environment they can better distinguish between what birds 'to eat' and 'not to eat' or – and this is an issue of greater importance to men – which birds to hunt and not to hunt.

Birds, animals or fish not regularly hunted or fished appear less frequently in conversation. Vassilikiots' comments about them are concise, comprising one or two stereotypical attributions. Here I present a few examples:

> 'Sharks are tasty, they can be caught', 'The flying fish (*Khelidhonopsaro*) is a fish with a tail and gills! It is edible (*trogetai*)', 'The bat (*Nykhteridha*) has breasts and she delivers babies like the goat. If you go close to where she keeps her young she can make you blind (*borei na se stravosei*)', 'The raven (*Korakas*) used to eat chicks and turkey-chicks. There are no ravens left nowadays, but we still say 'The place of the raven' [a place-name: *I Thesi tou Korakou*]'.[14]

Three different species of nocturnal birds of prey are recognised and can be named by most men and women in Vassilikos. These are the little owl (*Koukouvagia*), the eagle owl (*Boufos*), and the Scops owl (*Gionis*).[15/16] An old woman explained to me why *Gionis*, the Scops owl, produces the strange sound from which it takes its Greek name:

> Gionis is calling the name of his brother, Antonis. He killed Antonis by accident while they were working together in the fields. Ashamed to return home and face his mother, he kept wandering until late at night, crying out 'Antonis' in despair, and in the end, he became a bird. He is still calling Antoni, Antoni, Antoni, (the old woman imitated the voice of the bird) gioni, gioni, gioni!

Explanations of this type, referring to a particular bird or animal as 'once human' (*itane kapote anthropos*) and being transformed into the species in question, for one reason or another (God's punishment or a mother's curse), are widespread in rural Greece. I was surprised to find so little ethnographic material of this kind in Vassilikos, where most of the local inhabitants insisted that they did not remember those 'things any more.' In a similar way, I failed to record folk songs and laments that use images borrowed from the wild fauna (e.g. wild birds) (cf. Danforth 1982: 62–5, 112–5).

Neither wolves, nor foxes – the latter being plentiful on mainland Greece – live on Zakynthos. In Vassilikos they exist only as protagonists in fairy tales, which I managed to persuade a couple of elderly women to recount. Young and older Vassilikiots dismiss those tales as obsolete and maintain that 'nowadays, no one bothers to tell such stories to children'. Those stories I have managed to record reflect allegorically on the working partnerships between farmers and on the division of labour between men and women[17], but, to my disappointment, they do not reveal much about local perceptions of wild animals. Furthermore, they diverge from the everyday, practical view of wolves and foxes as pests (cf. Marvin 2000b, Lindquist 2000) or creatures which are inherently wild, harmful and violent (cf. Moore 1994). The following description of the wolf by one local farmer illustrates this point further:

> The wolf is a greedy (*aplisto*) animal. When he gets in a flock of sheep he kills a hundred and one sheep until he bursts (*mekhrei na skasei*)... The wolf catches the donkey with the greatest ease in the world (*me tin megalyteri efkolia tou kosmou*). He lies down on his back. The donkey goes to see out of curiosity and the wolf grabs the donkey by the nose.

Wild animals and orderly farms

Vassilikiots feel particularly fortunate about the absence of foxes and other large mammal predators from their island. This is why they let poultry, and sometimes rabbits, roam freely around the farmstead in search of food. Sometimes however, those chickens and rabbits fall victim to smaller predators – martens, hedgehogs and large rats – much to the dismay of their owners-cum-carers. 'Martens and hedgehogs take small animals from their nests and cause us damage (*zimia*)', the farmers complain, unable to hide the frustration and anger triggered by such unpredictable circumstances. The death of a domestic animal 'in the teeth' of a wild predator is always a cause of sorrow (*stenokhoria*), a sentiment expressed and shared in contexts such as the one described here:

> One early winter afternoon, I was enjoying my coffee and some preserved fruit along with two women, the hostess and her neighbour, when the husband of the former returned home. He immediately started complaining about some rats that had caused the death of two young rabbits. He discovered their dead bodies in their cages that morning. His wife vividly portrayed the grief (*stenokhoria*) of the mother rabbit, while the other woman and the farmer started making assumptions about where the rats came from. The nearest wood (*logos*) was unanimously declared by the company as the most probable habitat of the rats, since both women had seen them disappearing in that direction. A vivid and detailed description of rat attacks followed, allowing me to visualise the day when one of the two women had seen a rat 'with its frightening

teeth.' She ambushed (*paramonepse*) it and scalded it with a dish full of hot water. The husband had a similar story to tell about the day he realised that a rat was coming every morning to the same spot, causing great disturbance among the chickens. Apparently, he watched for the rat (*tou estise karteri*) with his gun and shot it. The grief the company felt on account of the rat attacks was evident. 'It's not that I care about the loss of one or two chickens [i.e. their monetary value],' one of the two women explained to me, 'but I am upset (*stenokhoriemai*) that I lost them.' When animals die in this way, everybody agreed that afternoon, their 'efforts go in vain.'

Vassilikiots repeatedly stress that they are not overly concerned with the monetary value of lost animals. Their grief, they claim, is mostly over the fact that the labour they have invested in caring for the young chicks and rabbits is dissipated. Feeding them, ensuring that young chick's food is not consumed by older chickens, collecting them every afternoon into crates or small cages to protect them from attacks, are some of the tasks that comprise this daily labour. To confine the young and active chicks into the hencoop is not an easy task, especially for older farmers, who feel justified considering all this work a struggle (*agona*). When the rats succeed in killing their animals Vassilikiots feel particularly disappointed. They claim that their efforts are not adequately rewarded and that their work or 'struggle' is wasted or 'lost'. 'It isn't worth so much effort,' they comment with resentment, 'it's a pity, (it takes) so much effort and all you get is the grief for the lost animal (*krima, tosos kopos, kai ti menei? I stenokhoria gia to khameno zoo*).'

In most conversations, the Vassilikiots' references to wasted or lost labour is not radically distinguished from the sorrow they feel for the lost animal. In fact, it is in terms of the care and labour spent on the particular animals that any sentiments or affection are articulated. Care and affection for an animal are indispensable parts of the greater system of order in the farm, one that entails punishment, discipline, security from predators and, eventually, the death of the animal itself. And there is always a point in time when the life of a domestic animal is terminated by its owners. In this case, the eventual and planned death of the animal is understood as part of a larger scale system of services offered by all members of the farm in the context of sustaining and maintaining the order of this complex microcosm. With its death the animal contributes to the general welfare of the farm, thus reciprocating the care it received in the past.

On the other hand, the sudden and unpredictable death of a domestic animal causes grief and feelings of disempowerment. Losing an animal to a wild predator or to a greedy landlord[18] constitutes a severe and premature interruption of the process of care, and it is thus inconsistent with the order of the farm and its principles. Not surprisingly, the dead or the appropriated animal is considered as having been 'lost' (*khameno*), for all the care and labour invested in it is indeed lost.

To protect their domestic animals – and all the care they have invested in them – from wild predators, Vassilikiots do not hesitate to pursue any harmful creature they encounter with persistence and determination. They consider this task an obligation to their households, but also a public duty, and feel particularly proud after successfully exterminating wild or dangerous animal intruders (cf. Knight 2000: 7). This is how a farmer in Vassilikos felt after pursuing a dangerous stray dog:

> One day I saw a *liariko* [piebald] dog attacking the goats of the Tzanetos family.[19] I shot once to make it get away from the goat, and with the second shot I wounded it (*to lavosa*) in the back. Somebody else found the same dog on a bench and he finished it off. I felt I was doing a service (*leitourgeima*) because I was protecting people's animals (*ta zoa ton anthropon*). This dog could do harm.[20]

Here, the farmer has not simply evaluated the wild animal according to the criterion of usefulness. The stray dog, like the ferret or the rat, was demonstrably harmful: 'an illegitimate killer' – to borrow an expression by Garry Marvin (2000b: 207, 208) – that has violated the order of the farm. The farmer's confrontation with it is only a small part of the ongoing struggle to establish and defend his or her position in the constantly changing, regenerating, and often threatening local environment.

Wild animals, however, are not simply seen as physical and notional enemies of the farm. They can often be subsumed – with various justifications – in the order of the farm and receive a fair amount of care despite their apparent uselessness. Such is the case of farmers who decide to keep captured wild animals alive, deviating from the local criteria of animal usefulness and the strict application of practical priorities. In their encounters with wild animals, Vassilikiots – as legitimate defenders of their households – reserve the right to punish, be merciful, or even exhibit care, and through care, affection. Here is a characteristic example:

> It was early February and I was in the fields of Vassilikos with Lefteris. We were collecting scattered animals on our usual route back to the farm, when suddenly he told me to 'stand still.' Right there, alas too close to my nimble friend, was a hare looking for cover in the thick grass. Without further delay, Lefteris seized it with his hand. He was holding the hare by the ears, the same way he usually holds his rabbits, only this time his face was shining with the excitement of success. The hare was triumphantly brought to the farm and put in a cage. If it were a female, Lefteris announced, he would allow it to mate with his male rabbits. The wild qualities of the hare, he explained to me, would revitalise the blood stock of his rabbits. His wife, however, thought it was best to eat it, because as she argued 'others will kill it, or it will die from sadness.' Both, however, remained there, looking at the beautiful hare with pride, amazement and admiration. The creature was displayed and commented upon for the whole afternoon and several days thereafter.

The story of the hare's capture became Lefteris's and mine – more his than mine – heroic tale for quite some time. My friend succeeded in realising the Greek proverb 'to catch a hare with one's hand,' reserved for people with acute alertness and high dexterity. The hare turned out to be male and was thus transformed into an enviable traditional *stifadho* [a local recipe], cooked with onions in red sauce by Lefteris's wife. 'It would get an unfriendly reception from the male rabbits of the farm and cause disturbance,' Lefteris rationalised *a posteriori*. Nevertheless, for some time thereafter, Lefteris and his wife were charmed by the hare's untamed qualities and seriously considered keeping it alive and putting it among the other animals on the farm. What is probably worth stressing at this point is the flexi-bility of the notion of usefulness that allows the farmer to invent, so to speak, a use for an animal that he or she is reluctant to kill. An alternative use, mating with the rabbits, was readily found in this case in order to justify and support the beau-tiful hare's right to life.

Another example that demonstrates the flexible application of the criteria that emphasise usefulness concerns a marten kept in a cage in Vassilikos. The marten (*Kounavi*)[21] is the largest predator on the island and poses a serious threat to free-ranging chickens. This was probably the reason why an elderly Vassilikiot woman had a whole collection of stuffed martens in her living room. 'My husband, like other men in the village, used to hunt martens all the time' she explained to me, 'they do harm to chickens.' Then, after a short pause she continued: 'There was a time, though, that one marten was caught in a snare (*dhokano*). We decided to keep it in a cage and it became tame. When we set it free again, it used to hang around my yard.' Listening to her story, I could not hide my disbelief. Neverthe-less, the Vassilikiot woman insisted that her story was true and further explained away her decision to keep a wild, and potentially harmful, animal on her farm by stating that: 'Martens kill snakes and rats.' She invented, thus, an alternative but fully legitimate function for an otherwise dangerous, but, in those circumstances, reformed creature.

The criteria of usefulness could be also twisted further to include decorative purposes, as often happens in Vassilikos with turtledoves. Turtledoves (*trygonia*)[22] are probably the most important game on the island. As a general rule the Vassi-likiots show particular ardour, zeal and devotion to hunting down all birds of this species that happen to fly over their heads – and the heads of unsuspecting tourists and visitors. Nevertheless, I noticed some cases when individuals kept tur-tledoves in large cages near or outside their houses. The cages were constructed from thin wire netting fitted on to large concrete bases painted with lime. The captured turtledoves, once slightly wounded by hunting guns, looked almost like canaries. Their keepers argued that since 'the birds fell into our hands (*sta kheria mas*) unable to fly but alive and in good condition, we let them stay on the farm for decorative purposes (*gia omorfia*).' The personal wish of the farmers to keep the wild birds alive was paired with a more prosaic 'use' or 'function', that of the pleasure to the eye. Over and above any other consideration, it would have been inappropriate for a farmer and passionate hunter to admit that he loved turtle-doves and appreciated their right to exist in any form other than barbecued. As the Vassilikiots say 'only a non-farmer and a city dweller would have argued so' and thus they make certain that the captured turtledoves serve at least some purpose, if only a decorative one.

The cases of the dolphin and the seal provide a similar final example. Both animals cause considerable destruction to fishing nets and the damage they produce is the same: big holes in the nets which are either mended with great difficulty or remain irreversibly damaged. The seals, however, are more frequently blamed for this destruction than the dolphins. Vassilikiots comment on the appearance and behaviour of the two animals in order to explain their different attitudes towards them. They maintain that 'the seal is ugly (*askhimi*), while the dolphin is an animal you look at with admiration (*to kamaroneis*).' Others recognise signs of 'friendly' behaviour exhibited by dolphins, which frequently approach and follow close behind fishing boats. An older respondent remembers:

> They [the people of the village] used to consider the dolphin as the most benign animal of the sea. It saves shipwrecked people (*navagous*). But at that time they didn't use fishing nets (*dhen rikhnan dhikhtya*)!... [so as to get angry about the damage caused to the nets].

Dolphins are portrayed as friendly, benign and beautiful. Seals are considered as 'less friendly' since they cannot be approached with the same ease. The 'social' portrait of the dolphin is contrasted with the 'wild' and 'distant' character of the seal, thus the local people express their sympathy for the dolphins. In addition, reference is made to some usefulness on the dolphin's part – that is, saving shipwrecked people – to further validate the Vassilikiots' preference for that animal. Here, the prevailing code of 'usefulness' is presented once again, but it is not the only criterion employed in determining the local people's evaluations.

The relationship of Vassilikiots to wild animals, as the examples above demonstrate, often manages to elude utilitarianism, albeit not always to evade the confines of a use-oriented discourse. Summoning a wide variety of criteria of usefulness, the farmers engage with the wild animal in a flexible relationship that entails competition and a constant evaluation of the animal's individual attributes. This is definitely a hierarchical relationship where the human partner is regarded as being legitimately the dominant one and rarely does she or he experience ambivalence regarding his or her position with respect to the animal or its fate. In the Vassilikiot cosmology human authority and superiority not only remains unchallenged, but it is also thought to be perceived and endorsed by the animal itself. Despite this neatly ordered outlook, however, wild animals, either as a result of their harmful potential to predators, their ability to deceive, or simply, by their beauty, succeed at times in gaining a precarious, but decisive advantage in their otherwise straightforward relationship with the human actor.

St Basil's classification of the animal world

In the detailed ethnography on the human-animal relationship presented so far, I have repeatedly referred to the local perception of human authority over non-human beings. This authority was most prominently expressed in Vassilikiot farmers' perceived entitlement to organise and impose 'order' on the farm-environment and their power to decide upon the fate of domesticated animals and captured wild animals. I have also implied that the Vassilikiots' perception of their dominance over animals and the natural world is supported, and actually reinforced, by an elaborate religious cosmology. Here, I will attempt to illustrate some principles underlining this cosmology, focusing on religious beliefs about non-human living beings and their role and position in the cosmology established by Greek Orthodox dogma.

To achieve this I have chosen to present in some detail St Basil's classification of non human beings as this becomes apparent in his *Homilies on the Hexaemeron*, a religious discourse composed of nine consecutive homilies delivered in Caesarea of Cappadocia around 370 AD.[23] This town was an important cultural and political centre in the Eastern Roman and Early Byzantine Empire. Most of the audience at whom the homilies were aimed would probably have been manual workers and farmers, listening to a homily in the morning before departing for work, and to a second in the evening as they returned home (Sakkos 1973, Papoutsopoulos 1992). The presence of some educated people in the audience can be inferred from some comments made by the author.

St Basil defines his primary objective in the *Hexaemeron* as an exegetical one. The nine homilies are an interpretation (*ermineia*) of the first chapter of Genesis, although human creation is excluded.[24] The author attempts to explain the meaning of Genesis in a way that will be comprehensible to a wider Christian public. Like most prominent Christian thinkers of his time, he was engaged in fighting heresies and establishing standards for the dogmatic interpretation of Holy Scripture. St Basil distinguishes sharply between his interpretation of Genesis and the work of pagan philosophers or heretics who apply allegorical interpretations to Holy Writ (*nomous alligorias, tropologiais*).[25] Being a man of learning, educated at the 'Philosophy School' of Athens, St Basil was well acquainted with the works of the ancient Greek scholars. In the nine homilies of the *Hexaemeron* he directly or indirectly alludes to Aristotle, Plutarch, Origen and others.[26] His knowledge of the extensive and systematic work of Aristotle on plants and animals is also apparent from the text. Indeed, St Basil even uses some of Aristotle's examples.

Unlike Aristotle who attempted to complete a systematic study of natural organisms[27], St Basil's primary intention is to praise the wisdom of the Creator, and show how divine Providence lies behind the diversity of living beings. However, despite his contempt for non-spiritually-oriented scholarship, St Basil's orderly description of living species entails a form of classification. In his description of fauna and flora, he explicitly and implicitly groups living beings into categories. Variation in animal and plant species is treated as the means of ordering his description and illustrating the meaning of Creation. In this process, peculiarities of individual species are used as the criteria for establishing variation among living organisms. Stability in variations between species in successive generations is understood as the perpetuation of 'order' in the universe, a form of 'order' introduced by the Creator at his command. In the remainder of this section I will present some information from homilies E, Z, H and Θ of the *Hexaemeron*, affording valuable insights into present-day knowledge, attitudes and popular beliefs regarding the natural world in Vassilikos.

Homily E is a discourse about the creation of plants. Plants were 'brought forth' out of the earth at God's command, St Basil explains, 'first the herb, then the trees.'[28] In three different parts of this homily, the author emphasises the correct order of the plant's generation and reproduction until the 'present time.' First there is germination, 'for, germination is the beginning of every herb and every plant.'[29] Then follows the generation of the 'green shoot', the seedling stage. Thirdly, the plant becomes 'a grass' or, in the case of the more complicated plants, the 'green foliage' is developed. In the final stage the fruit ripens and the 'perfection' of the seed is completed.[30]

Apart from the initial distinction between herbs and trees, which is directly implied by the text in Genesis, St Basil orders plants according to their use by people[31] such as the ability to bear fruit, suitability for building shelters or ships or the potential for use as fuel.[32] He also refers to the plants' decorative role, medical properties and their role as nutrition for animals.[33] 'There is not one plant without worth, not one without use', St Basil argues, 'either it provides food for some animal' or it serves as a medicine for people (E par. 20). Even in cases where plants are 'useful for the other living creatures', the author of the *Hexaemeron* maintains that 'the profit they receive passes over' to the human protagonists (E par. 5). The text of Genesis is his authority for asserting that the creation of plant life took place in order not only to meet the needs of herbivorous animals, but also to satisfy the needs of human beings.[34]

St Basil's description of plant species is guided by reference to their appearance and physical attributes. The shape and formation of roots, trunks and branches, as well as the shape, colour and flavour of the fruit or the formation of the foliage, are treated by St Basil as indicative of the variations among different species of plants.[35] The infinite magnitude of natural variation is interpreted by him as an illustration of divine wisdom. He further explains that nature (*fysis*), which in this context is synonymous with the divine order, provided plants with the appropriate characteristics and shapes, fitting them for survival. The functional character of plant structure is treated by St Basil as an illustration of divinely inspired order and causality. 'Nothing happens without cause', he clearly states, 'nothing by chance; all things involve a certain ineffable wisdom (E par. 46).'

If we consider plants as ornaments of the Earth, St Basil maintains in Homily Z, aquatic animals are ornaments of the waters: all forms of water – sea, rivers, lakes, even slime and ponds – became productive, at the divine command, producing all sorts of creatures that can swim (*plota, nikhtika*). These creatures, believed to be created by divine command in the element of water, are not merely fish. Frogs and mosquitoes like 'seals, crocodiles, hippopotamuses, crabs' are all considered to belong to the same general category.[36] St Basil clarifies this point: 'Even though some of the aquatic animals have feet and are able to walk, yet their ability to swim is antecedent (Z par. 3).'

More important than any other characteristic, the relationship of aquatic animals with water, the element they live in and were produced from, is the primary criterion for grouping those animals together. As the *Hexaemeron* clarifies: 'Every creature able to swim, whether it swims at the surface of the water or cuts through its depths, is of the nature of crawling creatures, since it makes its way through a body of water (Z par. 3).' In his next homily, St Basil will demonstrate the importance of 'crawling', as a method of moving in an element like water or air, for classifying swimming and flying animals in one general category. In homily Z, however, he is merely interested in establishing the relationship of aquatic animals to water. For this purpose, the author examines part of the fishes' internal structure, their organs for breathing.[37] He accurately contrasts the respiration of fish by the 'dilation and folding of the gills' with human respiration through the lungs and demonstrates why fish cannot remain alive away from water, the element from, and, for which they were created.[38]

The author proceeds in his orderly description of aquatic animals to discuss size, habitat, lifestyle, method of procreation and external characteristics of body structure or appearance. Aquatic animals are subsequently divided into those which live in the open and deep sea and those which live close to the shore, 'those which cling to rocks, those which travel in shoals, those which live solitary, the sea monsters, the enormous, and the tiniest fish (Z par. 5).' Aquatic beings which bear live offspring (*vivipara*), like sharks, dogfish, seals, dolphins, rays are grouped separately from those beings which, like most kinds of fish, produce eggs (*ovipara*). The latter category is further subdivided into 'scaly and horny scaled' fish, 'those which have fins and those which do not' (Z par. 4–8). St Basil also maintains that 'fish have a specific space to live in, a characteristic nature, a distinct feeding and a peculiar mode of life (Z par. 6).'

In the Greek translation of *Genesis* and the *Hexaemeron* the word genus (*genos*) is used instead of the words 'kind, species and class' used by the English translations (let the waters bring forth crawling creatures of different kinds = different genera).[39] The following categories of animals are termed genera by St Basil: testaceans (mussels, scallops, sea snails, conchs etc), crustaceans (crayfish, crabs etc), and soft fish (polyps, cuttlefish etc). The *ovipara* and *vivipara* (like most cetaceans) constitute different genera, in the same way that cetaceans (big aquatic animals) and tiny fish belong to separate genera.[40] According to St Basil, 'every genus has a particular name, food, shape, size and quality of flesh; all genera are distinguished by great differences and are divided into different species (Z par. 9).'

Appearance, mode of reproduction and behaviour, are indiscriminately used by St Basil as criteria for grouping aquatic animals into different genera. The author is not concerned with particular details leading to a systematic classification. He clarified this point in Homily E. He is categorical, however, when the distinction between different life forms is implied by the text of Genesis. For example, plants are not mixed with swimming or flying animals. However, when the grouping of living creatures is not directed by Holy Scripture, St Basil employs a variety of criteria to order his descriptive account of the various genera of animals. His purpose is to establish the distinctions between different categories of aquatic animals and ensure that the characteristics of each category remain unchanged through generations.[41]

In Homily Z, St Basil states that aquatic creatures are the first beings in the Creation to possess 'life and sensation'. The author sharply contrasts aquatic animals with plants: 'plants and trees, even if they are said to live because they share the power of nourishing themselves and of growing, yet are not animals nor are they animate (Z par. 3).' This is the first basic distinction between animal species drawn by St Basil, the one between inanimate plants and animate beings. Aquatic animals, are animate beings, but according to St Basil's interpretation their life is in some sense imperfect; they lack the ability to 'speak or reason', 'be tamed' or 'endure the touch of the human hand'. Using the example of fish migration, the author illustrates that since aquatic animals 'do not have reason of their own...they have the law of nature strongly established which shows what must be done' (Z par. 6–22). With these words the author anticipates his subsequent distinction between different orders of animate beings.

In the next homily (H) the author of the *Hexaemeron* offers more information concerning the spiritual state of animals. He begins by comparing the lives of swimming creatures and animals of the land. According to the text of Genesis, aquatic animals *have life*, while animals of the earth are *living creatures*. This distinction renders the animals of the earth superior. The author discusses this in detail stating that aquatic animals have a rather imperfect life, since 'they live in the dense element of water.' He demonstrates this point by referring to the limitations of their senses: their hearing is poor, their sight is dim, they are unable to remember, imagine and recognise the familiar. Because of their limited perception, St Basil infers that among the aquatic beings, the life of the flesh directs the motives of the soul.[42] The author describes fish as creatures which are 'voiceless, but also incapable of being tamed or taught or trained for any participation in the life of humans (H par. 4–5).'

By contrast, St Basil argues, the life of land animals is *more perfect* and for this reason their soul has hegemony over the body. The sensations of the land animals are more acute. Most quadrupeds perceive the events happening in the present time with acuity and remember past events with precision. This is why, St Basil concludes, in the case of land animals it was commanded [by God] that a soul be created that would shape the body. The animals that live on the land possess a stronger life force. For St Basil, although land animals lack the power of reason – this is treated as an undisputed fact – they have a voice and can express sentiments with it. They express happiness and sadness, and recognition and hunger and numerous other states, which St Basil calls emotions – and a behaviourist psychologist would call drives. All these arguments are used by the author of the *Hexaemeron* to demonstrate the superiority of land animals over aquatic creatures.[43]

While St Basil is examining the animals of the land in Homily H he realises that he has completely omitted one of the three parts of living creation, the flying creatures. After apologising for his mistake he immediately proceeds to discuss the creatures of the air (*ptina*), starting from a comparison between them and the marine animals (*plota*). Both 'cut' or 'move forward through' an ethereal or liquid medium like water or air, assisted by their tails, fins or wings. This ability comes out of their common origin, St Basil maintains (H par. 11).' However, there is a difference between birds and fish because 'none of the winged creatures is without feet.' Feet were given to birds in order for them to subsist, since they find nourishment on earth.[44]

The author of the *Hexaemeron*, faithful to the text of Genesis, presents flying creatures as deriving from the waters, like aquatic animals. Although he does not explicitly compare flying creatures with animals of the land, one may suppose that the former are inferior to the latter for the reasons already stated in the comparison between aquatic and land creatures (both swimming and flying creatures came out of water and 'have life' but are not 'living creatures').

It is worth mentioning that insects and birds are incorporated into the same general category, the flying creatures (*ptina*). This does not mean that St Basil is ignorant of the structural differences between birds and insects. At some point in his homily he explains that creatures like bees and wasps are called 'insects', because 'they appear cut into segments all around', as the Greek etymology of their name denotes [*entoma*]. Moreover, he explains that insects do not breathe, nor do they have lungs but they absorb the air through all parts of their bodies.[45]

St Basil orders his description of the flying creatures by reference to criteria such as nourishment, physical appearance, mode of life and group organisation. He divides the flying creatures into the 'genera' of carnivora, seed-picking and omnivorous birds and explains that their physical structure is analogous to the food they eat and the kind of life they have. Among the omnivorous birds, he argues, there are several subdivisions. Some birds prefer to live in flocks, others have chosen a collective form of life.[46] Among the latter, some are autonomous, without any superiors, while some others accept the command or headship of a leader. St Basil states that more variation can be found in the former category, since some birds are permanent residents of a particular place and others migrate to distant lands before winter.[47]

One, final distinction drawn by St Basil among the flying creatures, is between the nocturnal genera of birds (*ta nykhterovia geni ton ornithon*) and those which 'fly about in the light of the day.'[48] In the former category he includes bats, owls, the nightingale, and night ravens. He remarks on the peculiarity of the bat, which is both a quadruped and a flying creature (*ptino*). The bat, St Basil states, is the only bird to use teeth, bear live offspring and fly in the air, not by use of feathered wings, but by means of a skin membrane.[49]

St Basil concludes his discourse on flying creatures with a lengthy discussion on the attributes and character of various birds. Parallels and metaphors are drawn out of the lives of these flying creatures for the purpose of making the audience contemplate moral qualities or values. This practice is employed by St Basil in all the homilies that I have examined, but the attribution of anthropomorphic characteristics becomes more frequent in the discussion of flying creatures and culminates in the description of land animals. 'Some irrational creatures are like members of a state' (*esti dhe tina kai politika ton alogon*), he comments, in an example about the organisation of bees (H par. 17). 'The conduct of the storks is not far from reasoning intelligence', St Basil argues, and commends their care of the aged members of their species (H par. 23–4). Similarly he praises the responsibility and orderly flight of the cranes (H par. 22), the companionship of bats (H par. 34), the vigilance of geese (they once saved the imperial city of Rome!) (H par. 36), and the love of the crow for its offspring (H par. 30), to mention but a few of St Basil's lively examples.

Several more anthropomorphic examples are mentioned by St Basil in his Θ homily on 'land animals', the last homily of the *Hexaemeron*. The author refers to the firmness of the ox, the sluggishness of the donkey, the horse's 'burning desire for the mate', the untamed nature of the wolf, the deceitfulness of the fox, the timid character of the deer, the industrious traits of the ant, and the gratitude and faithfulness of the dog.[50] St Basil maintains that each animal at its creation received a distinctive natural property or virtue (*fysikon idhioma*). Along with the lion, for example, was brought forth (born) the lion's anger, the lion's pride, and its solitary and unsocial mode of life. Additionally, St Basil maintains that the bodies of the animals were created as analogies of the innate characteristics of their souls (*tis psykhis kinimasi synepomenon to soma*). For example, the leopard was given an agile and light body, suitable to realise the urges of its soul. The bear received a stiff, heavy, not distinctly articulated body, which resembles its lazy, insidious and secretive character.[51]

The distinction between land animals and human beings is very important for St Basil. 'The beasts are earthy and they watch towards the earth', he declares.[52] Human superiority in 'the value of the soul' is evident in the construction of the body. The etymology of the Greek word *anthropos* – *ano throsko*: I look/watch upwards – is indicative of St Basil's argument. Human heads 'stand erect toward the heavens', human eyes 'look upward', the author states rhetorically. Similarly, the configuration of 'quadruped' animals signifies their close relation with the earth. The author observes: 'their head bends toward the earth and looks toward their belly and pursues its pleasure in every way (Θ par. 8).'

St Basil maintains that land animals have one kind of soul, characterised by lack of reason (*alogia*).[53] They differ from each other however, in terms of distinct properties or virtues, like the anthropomorphic ones I have already described. St Basil maintains that God compensated the land animal's lack of reason by providing them with superior sensory abilities and further demonstrates his point with examples.[54] The lamb can recognise its mother's bleating among countless other identical sheep due to a form of perception which is more acute than the human one.[55] The dog, too, another animal without reason, has sensory facilities equivalent to reason, claims St Basil in another example. When the dog is following the tracks of a wild beast and examines various possible routes, it locates the correct way by a process of elimination. The dog was taught by nature, what 'so-called' wise people discovered with much difficulty by drawing lines in the dust, notes St Basil, using the opportunity to speak ironically of the pagan philosophers and mathematicians.[56]

Contemplating the creation of the natural world, St Basil anticipates some elementary observations of modern ecology; he recognises that those animals that are easily captured reproduce more rapidly. By contrast, predators like the lion have very few offspring.[57] But for St Basil, all manifestations of the creation show the wisdom of the Creator. Divine Providence did not deprive any being of what was 'necessary' or 'useful' for its survival, nor add anything 'superfluous' or 'unnecessary'.[58] The author demonstrates this idea by examining the body structure of animals, in a fashion reminiscent of Lamarck:

> The camel's neck is long in order that it may be brought to the level of his feet and he may reach the grass on which he lives. The bear's neck and also that of the lion, tiger, and the other animals of the family, is short and is buried in the shoulders, because their nourishment does not come from grass and they do not have to bend down to the ground (Θ par. 24, translation by Sister Agnes Clare Way 1963: 144).

Nobody can accuse the Creator of creating animals which are poisonous, destructive and hostile to human life, St Basil maintains. To do so, would have been like accusing a pedagogue of putting delinquent youth 'in order' by means of punishment ('rods and whips!'). The author, throughout the homilies of the *Hexaemeron*, consistently supports the idea that dangerous or harmful creatures serve to educate people and test the power of their faith.[59] For St Basil the creation of animals, like the creation of plants, has a non-random, intentional character. The features of individual species are designed by a divine source in order to fulfil a two-fold purpose: to facilitate and perpetuate the life functions of the particular species and simultaneously benefit, directly or indirectly, 'mankind'. I will conclude this section with an extract from St Basil's description of the elephant, where one can observe those two kinds of causality, based respectively on a functional and an anthropocentric logic:

But what is the reason for the elephant's trunk? Because the huge creature, the largest of land animals, produced for the consternation of those encountering it, had to have a very fleshy and massive body. If an immense neck proportionate to his legs had been given to this animal, it would have been hard to manage, since it would always be falling down because of its excessive weight. As it is, however, his head is attached to his backbone by a few vertebrae of the neck and he has the trunk which fulfils the function of the neck and through which he procures nourishment for himself and draws up water.

... As we have said, the trunk, which is serpent-like and rather flexible by nature, carries the food up from the ground. Thus the statement is true that nothing superfluous or lacking can be found in creation. Yet, this animal, which is so immense in size, God has made subject to us so that, when taught, it understands, and when struck, it submits. By this He clearly teaches that He has placed all things under us because we have been made in the image of the Creator (Θ par. 25–8, translation by Sister Agnes Clare Way 1963: 144–5).

The *Hexaemeron* and the hierarchy of species

Taxonomic inquiry is often associated with Mary Douglas, Edmund Leach, taboos, pollution and prohibitions (Douglas 1957, 1966, 1975; Leach 1964, 1969). Both anthropologists, during the 1960s, approached animal classification from a similar perspective.[60] In their work, different animal categories operate as units of 'order', the respective boundaries of which are charged with pollution, negative prohibitions, or even extremely positive, almost sacred, associations. Both authors concentrate on the powerful conjunctions of diverse categories, the instances in which particular animals fit criteria defining separate categories. What I find interesting in this form of analysis is not the apparent preoccupation with anomalies, but the idea of 'order' itself: how different levels of 'distance' from the human self – to use a schema applied by Leach (1964) – reflect the order of relations between different categories of animal species and human beings. Animal classification defines an 'order' of hierarchies and priorities between living beings, in which the human self holds a dominant position.

In Leach's and Douglas's work, the idea of 'order' appears to be a central concept in understanding systems of animal classification, as it is in St Basil's *Hexaemeron*. The verb 'to classify' is almost synonymous with the verb 'to order'; in Greek, the equivalent verb is *taxinomo*, where *taxi* means order [from which we get the term 'taxonomy']. But if 'order' for St Basil is synonymous with the 'divine order', for Douglas and Leach 'order' is something similar: it is primarily 'social order'. And in as much as 'divine order' in St Basil's interpretation is a basic assumption rather than a mere methodological tool, several well-known anthropological studies in the 1960s treat 'social order' as an animated entity embodying classification.

Lévi-Strauss's structuralism was immensely influential in animal classification studies produced during this period, although some anthropologists, such as Mary Douglas, would deny Lévi-Strauss's ascendancy. What is of direct interest to my work is that the concept of 'order' is equally important in the work of the French anthropologist. Reading 'order', or identifying 'order', in classificatory systems is elevated into something greater than a simple prerequisite for establishing the existence of structures. In *The Savage Mind* (1962), the concept is developed into a 'demand for order', it becomes an underlying principle of the human mind. Classification does not simply reflect the structuring of social relations; it is the product of the human mind's need for order. This allows more space for human agency: for Lévi-Strauss in *The Savage Mind*, the stimulus structuring classification is not social order, but human beings attempting to make sense of their environment.

For Lévi-Strauss, the dynamic character of the concept 'species' is dependent upon the structural tensions between opposing categories. A species of animal has something to tell us, but only if it is placed against a definitional background of other species (Lévi-Strauss 1962: 136). St Basil's homilies do not acknowledge this kind of argument. In the *Hexaemeron*, different species or genera of animals acquire meaning independent of their given relationship with other species or human beings. Their relational value is predetermined by well-established religious hierarchies and priorities; meaning is ascribed to them at the very moment their position in the cosmological hierarchy is defined. The emphasis given to dichotomies and oppositions between different categories (Lévi-Strauss), or mediators (Leach), or anomalies (Douglas), provide little help for my analysis of the *Hexaemeron*, where I am directly concerned with the relationships between different orders of animals, and the relationship between animals and people.

In the homilies of the *Hexaemeron* human beings are not defined in terms of animals, nor animals in terms of human beings. A comparison of this sort would have been unthinkable for St Basil, or my contemporary informants on Zakynthos. The oppositions between man and animals, and between animals and inanimate organisms (plants), which are clearly expressed and stressed in the *Hexaemeron*, are defined in terms of an anthropocentric perspective superimposing predetermined hierarchies. This kind of classificatory logic represents levels of distancing the self from other natural categories – the conceptual schema applied by Leach (1964) and Tambiah (1969) – but on a vertical axis, where superiority or inferiority is taken for granted, being established from the start in a rather self-conscious fashion. Tambiah describes Lévi-Strauss as 'using natural models of differentiation to express social relations' (1969: 165). St Basil does the reverse: he consciously applies a theocentric model of differentiation in order to account for natural relations.

In Bulmer's essay 'Why the Cassowary is not a Bird' (1967), there is an interesting, short discussion of the criteria by which the Karam classify animals. I consider this, compared to Bulmer's larger concerns with the cassowary and the preoccupation of his time with anomalies, to be a more constructive approach to animal classification. Bulmer discusses the 'broadest groupings' and 'smallest units' in Karam taxonomy, interested in the logic permeating these two levels of classification. Furthermore, he observes that, at the lower small-scale taxonomic level, classification is based on a detailed, highly accurate knowledge of natural history comparable with the observations of the 'scientific zoologist'.[61] Those objective biological criteria, however, lose their relative importance at the broadest, upper end of the scale of categorisation, where classification is determined by cultural priorities. Bulmer's observations can be further expanded to cover the animal classification in the *Hexaemeron*.

Morphological characteristics, behavioural patterns, means of procreation, habitat, nutrition, and lifestyle are all criteria employed by St Basil in his categorisation of different species into genera.[62] All these criteria are used interchangeably to group species according to common properties. If the categories thus defined overlap, it is not of any particular significance for St Basil. The notion of 'order' employed by him is not threatened by minor inconsistencies of this kind. Since Holy Scripture does not provide any definite criteria for such a categorisation, the religious scholar applies a broad range of classificatory criteria based on contemporary empirical knowledge. Examples are drawn even from Aristotle, whose categorisation for the sake of systematisation is anathema to St Basil.

Categorisation according to genera in the *Hexaemeron* does not affect the implicit hierarchy between animate and inanimate beings. Furthermore, it fails to offer suitable ground for moral precepts. It is not surprising therefore, that St Basil treats this level of classification as being relatively insignificant. For him it is important to demonstrate that all species occupy a place in creation and reproduce themselves in a way that preserves the identity of their 'kind', as is stated in Genesis.

In contrast with the lower end of the scale of classification, the initial distinctions between animate beings are explicitly defined in Genesis. Three major categories of aquatic, 'flying' and land creatures have been recognised as classificatory categories in the anthropological literature by Douglas (1975: 263–5) and Leach (1969). St Basil offers more information, from the point of view of a faithful Christian and a dogmatic theologian. Aquatic and flying creatures, for example, are presented as having a common ancestry in the water, and are a form of life, which is somewhat 'imperfect'. The way these creatures move their bodies in a medium like water or air – flying is presented as analogous to swimming – is used by St Basil as a standard for establishing their identity. Land animals were 'brought forth' out of the earth and are portrayed as superior to aquatic and flying creatures, yet demonstrably inferior to human beings. The motives of their soul, like the construction and origin of their physical body, are described as being 'earthy'.

Plants, like land animals, were 'brought forth' out of the earth. But plants are believed to be inferior organisms, not even considered 'animate' beings. In fact, the phrase *plants are inferior organisms*, reflects my own perception of plants as organisms, not a judgement made by the author of the *Hexaemeron*. For St Basil, plants are simply 'inanimate'. They belong to a different, inferior order; this is why the process of categorising animate beings in the *Hexaemeron* begins with the distinction between aquatic, flying and land animals. The following diagram portraits the association of physical elements, with respective categories of animals, as well as the vertical hierarchy of their respective states of life, as expressed in the *Hexaemeron*.

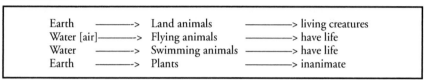

Animal classification in the *Hexaemeron*, reveals an implicit hierarchy between organisms of different orders, occupying different space and having different roles in the universe. The above diagram is made complete by the addition of human beings at the apex of the hierarchy, since it is in relation to the human social self that the hierarchy is made meaningful. In the following diagram, horizontal lines separate categories of absolute boundaries, represented by the distinctions between plants and animate beings or between human beings and 'beings with no reason'. The addition of an extra absolute dividing line, between the Creator of the universe (in triadic form) and the created beings, concludes this schematic representation of the cosmology in the *Hexaemeron*.

Heaven <——— Creator, in triadic form.		
Earth ——-> Human beings ——-> made in the image of the Creator		
Earth ——-> Land animals ——-> living creatures Water [air] ——-> Flying animals ——-> have life Water ——-> Swimming animals ——-> have life		
Earth ——-> Plants ——-> inanimate		

In St Basil's homilies, the relationship of plants and animals to a physical medium or element like earth, water or air operates as a primary conceptual association, which directly informs their categorisation and place in a hierarchy of relationships. Comparing animal classification in the *Hexaemeron* with my own ethnographic experience on Zakynthos, I notice that similar classificatory criteria operate in both cases. The exercise of defining primary categories of animal classification according to media or elements 'on' or 'in' which different categories of animals live, is a commonplace classificatory strategy employed by the farmers in Vassilikos. Vassilikiot farmers, in their oral accounts of the local fauna, employ identical distinctions between sea, land and flying creatures. This observation does not imply that Vassilikiots consider seals and sea turtles as fish. Rather, it suggests that a form of animal classification based on the animal's habitat is a convenient, practical strategy by which rural Greeks describe animals in a given environment and locate themselves within it.

Conclusion

The relationship of Vassilikiots to wild animals, as expressed by the Vassilikiots themselves, is a one-way relationship. Vassilikiots perceive non-domesticated animals in terms of their own established presence in the local environment. They refer to wild animals in relation to their own point of view, their position as guardians of welfare and order on their farms. They are concerned about the potential 'harm' (*zimia*) or 'use' (*khrisimotita*) wild animals may 'cause' to (*kanoun*), or 'have' for (*ehoun*) their own households, that is, themselves and all the domesticated animals and plants on their farm. Their attitudes towards wild animals are expressed in accordance with criteria focusing on the usefulness of the animal in question to the human household and its animate or inanimate constituents, and usually follow three general tendencies. First, lack of benefit or harm done by the wild animal results in indifference. Second, the edibility of a wild animal renders it a legitimate target for hunting – a positive characteristic – and justifies its predation. Since hunting is, in general, celebrated in the narratives of the local people, Vassilikiots are eager to talk about the 'huntable' animals and share their knowledge and experience of hunting them. Third, animals locally portrayed as causing 'harm or damage' (*zimia*) are pursued with anger and resentment. Harmful animals are an obvious threat to the farmer's persistent efforts to establish a form of 'order' in the farm environment.

Predation by wild animals on domestic animals arouses sentiments of distress (*stenokhoria*) and anger (*thymos*) in their owners and caretakers (cf. Moore 1994: 83, 86). The process of 'caring' (*frontidha*) is interrupted and a significant amount of effort and labour is 'lost' (*khanetai*) along with the dead animals. Vassilikiot farmers express their disappointment at these unpredictable circumstances in ways similar to their reactions to natural calamities (e.g. bad weather or epidemics). In practice, however, they do not confine themselves to pessimistic statements but actively protect the animals of the farm from intruders by using guns, poison or other means. In this case, 'the illegitimate killer becomes an object of legitimate killing' (Marvin 2000b: 208), while the farmers' antagonism to the wild side of the physical environment is expressed in a direct and explicit form.

A wild animal's potential usefulness (or lack of it) is a fundamental consideration in Vassilikiots' evaluations of animals and informs their relationship with them directly. However, despite this general attitude, the villagers do not always apply a strictly utilitarian approach towards wild animals. Although they would normally kill 'harmful' (*vlavera*) animals whenever possible, I recorded a few cases in which the farmers kept wild animals in captivity and/or allowed them to remain alive. In those cases, characteristics of wild animals other than their practical 'use', such as their beauty or their friendly behaviour, were the rationale for keeping them on the farm. However, unlike city dwellers or environmentalists, Vassilikiots never justify their protectionist attitudes towards wild animals in terms of affection. Instead, the villagers would think of alternative forms of 'use or function' to rationalise their non-utilitarian decisions. Rationalisations of this kind reflect people's concern to be consistent with the criteria of usefulness they have already defined, but at the same time indicate their personal freedom to negotiate their relationship with wild animals and apply their personal decisions at a practical level.

Local beliefs that inform the relationship between people and wild animals in Vassilikos are also consistent with another idea: the axiom of human authority over physical organisms of all kinds. Without hesitation, Vassilikiots exercise their perceived right to decide upon the fate of every wild animal they encounter. They feel absolutely confident in applying their own personal conceptions of order and justice to all creatures found in the physical environment and especially those wild ones which are not already subjected to the care and order of the farm. This confident, guiding attitude of the Vassilikiots towards wild animals is in accordance with their religious beliefs, which justifies a conceptualisation of the indigenous self as the guardian or caretaker of the natural world. Their religious cosmology portrays human beings as having the authority – the biblical 'dominion' – to utilise physical resources for the benefit of their households. According to this view animals and plants are created by God in relation to man and for man's benefit.

To further illustrate the religious cultural tradition that informs Vassilikiot beliefs towards animals, I have discussed some central themes in a particularly influential theological discourse, St Basil's *Homilies on the Hexaemeron*. The work of St Basil does not merely comprise an interpretation of Genesis; it is an interpretation of the natural world according to the criteria established by Genesis. When the author systematically examines the characteristics of animals or plants he 'sees' proof of divine causality. Religious faith and the text of Genesis provide the primary classificatory principles, a kind of model according to which an understanding of the natural world is constructed. St Basil organises his material with the intention of identifying the underlying 'order' of the natural world. He responds to a given 'demand for order' – to facilitate his audience's understanding, to establish dogmatically a correct way of perceiving natural creation – but he responds to this 'demand for order' consciously and purposefully. Here, we have a case of structuring classification according to a given socially defined system of 'order', in a way which is too deliberate and too conscious to fit either Lévi-Strauss's (1966) or Durkheim and Mauss's (1963) structuralist model.

St Basil's primary classificatory principles are provided by the Bible and are taken for granted by the author, when he refers to the higher more inclusive classificatory categories. The resulting form of categorisation, which is culturally prescribed, may seem irrelevant to the empirically oriented naturalist. It was, however, historically relevant for the audience listening to the *Hexaemeron* and appears to be accepted without much hesitation by my interlocutors in Vassilikos. However, at the lower, less inclusive level of animal categorisation, St Basil's discourse is dramatically emancipated from the religious constraints that bind the initial conceptual dichotomies between creatures of the sea, air and earth. This is in accordance with Scot Atran's observation that 'basic level' taxonomic categorisation is founded on 'absolute' knowledge, grounded in empirical reality rather than cultural considerations (Atran 1990: 214, 5–6, 29, 56 1993: 57–9, 64; see also Ellen 1993: 93–4); the same point was made by Bulmer (1967: 6, 1970: 1072–3) some twenty years earlier. The multiple criteria shaping St Basil's orderly description of the animal world at this 'lowest' or most 'basic' level depend on animal morphology and behaviour, as well as a wide array of folk-zoological information and beliefs. It is here that Aristotle's naturalistic-empirical observations appear in St Basil's text, despite the latter's implicit antipathy for the former, which culminates in the deliberate avoidance of mentioning Aristotle by name.[63]

Animal classification in respect of physical elements like earth, air and water has deep roots within a particular ethnographic and historical context. The tendency to attribute special significance to elements of this kind, was characteristic of a long tradition of ancient Greek philosophers and scholars. The synthesis[64] of contemporary folk and natural history with Christian ideas is evident in the homilies of the *Hexaemeron*. As I have already mentioned, St Basil consistently employs folk understandings of natural history to fill the taxonomic gaps in the religious cosmology, especially at the lower level of classification. Charles Stewart remarks that 'synthetic religions, such as Greek Orthodoxy, ...traversed a period of active syncretism in the past but have now emerged as unified theological structures' (1991: 7). Present day Vassilikiot farmers, being for all practical purposes unaware of the historical processes of religious synthesis in the past, face their local natural environment fully equipped with a coherent religious cosmology that guarantees their right to dominance and authority over non-human beings. Their understanding of the human-animal relationship parallels the hierarchies identified in the *Hexaemeron*, and their general attitude towards the physical environment is indicative of a well-established anthropocentric tradition.

In St Basil's classification system, as much as in my Vassilikiot respondents' everyday discourse, use-oriented practical evaluations of animals exist side by side with morphological descriptions. Criteria based on usefulness consistently shape the Vassilikiots' understanding of non-human beings, while in St Basil's homilies, animals are presented as serving to bring benefits to man. Even particular animal characteristics, morphological and behavioural, are understood as serving, directly or indirectly, 'mankind', because, as is plainly stated by St Basil, all beings created by God are useful. This is why St Basil repeatedly argues that even useless and dangerous animals serve a function: they teach men moral lessons. What Berlin (1988, 1992; Berlin & Berlin 1983) would have called 'a utilitarian' explanation is superimposed here on the perceptually recognisable physical reality.[65/66] In fact, the culturally determined explanation embraces the practical use-oriented one and the perceptual recognition merges, as secondary supportive evidence, with antecedent well-established anthropocentric priorities and hierarchies.

This culturally specific anthropocentric perspective with respect to natural organisms is expressed both at the theological level of reasoning and in the everyday discourse and practice of my respondents in Vassilikos. It constitutes a coherent, pragmatic approach towards the physical world, which had remained virtually unchallenged until the recent appearance of environmentalists and conservationists on the island of Zakynthos.[67] The environmentalists exercise pressure on the state authorities to enforce the conservation of endangered species, such as the Loggerhead turtles, the Mediterranean monk-seals and a few species of birds, such as the turtledoves which are threatened by unrestrained hunting. In their campaigns, they emphasise the uniqueness of wild animals as independent creatures participating in an interdependent natural ecosystem. According to this view, turtles or seals have an inalienable right to exist in nature, sharing its resources with human beings. To ensure the endangered species' survival, the environmentalists demand constraints on the human population and the Vassilikiots' activities in the local environment. But as this chapter has made clear, the priorities of the environmentalists and the Vassilikiot farmers do not coincide. For the Vassilikiots, wild animals, such as the ones the environmentalists attempt to protect, occupy a peripheral position in the natural environment; their existence is defined in terms of the indigenous self's established presence on the land and the welfare of farming households. To prioritise the perceived needs of neighbouring fauna would seem to most Vassilikiots not only ludicrous, but a perversion of what they define as the natural order of things.

NOTES

1 The Vassilikiots' remarkable passion for hunting will be examined in Chapter Eight.
2 In particular, two other 'Holy Fathers' of the same era, St Gregory of Nyssa and St Gregory of Nazianzus.
3 The robin, (*Kokkinolaimis*), *Erithacus rubecula*, is called *Tsipourdhelos* in Zakynthos.

4 Robins do migrate but to far more northerly destinations.

5 '*Se ti khrisimevei i khelona?* or '*poia einai i khrisimotis tis khelonas?*.

6 Loggerhead Sea Turtle, *Caretta caretta.*

7 The general efforts of the environmentalists to 'educate' the public about the necessity of turtle conservation (through various leaflets, information kiosks, etc).

8 '*Dhen kanoun kanena kako, itan ligaki khrisimes tha borouse na pei kaneis ... ta ayga tous itan trofi gia ta skylia ...*'

9 The Mediterranean Monk Seal, *Monachus monachus.*

10 Having conducted fieldwork in Alonnissos, another Greek island, where seals exist in larger numbers and the local people depend on fishing to a greater extent than on Zakynthos, I recorded similar accusations about the seal. Like my Zakynthian informants, the people of Alonnissos emphasized the damage caused to their fishing nets by seals. The fishermen admit that they often had to shoot them before the passing of the conservation law, while some older men could remember that 'in the past people were using the seal's fat for lighting and the seal's skin for making rustic shoes (*tsaroukhia*)'.

11 Peregrine, (*Petritis*), *Falco peregrinus.* Sparrowhawk, (*Xefteri*), *Accipiter nisus.* Goshawk, (*Dhiplosaino*), *Accipiter gentilis* (in Zakynthos called *Barbouni*).

12 Lesser Kestrel, (*Vrakhokirkinezo*), *Falco tinnunculus* (in Zakynthos simply referred to as *Kirkinezi*). Black Kite, (*Tsiftis*), *Milvus migrans* (in Zakynthos called *Loukaina*).

13 One of the few brief comments Vassilikiots made about the Black Kite was: 'The *Loukaina* eats sick chickens'.

14 Raven, (*Korakas*), *Corvus corax.*

15 Little Owl, (*Koukouvagia*), *Athene noctua.* Eagle Owl, (*Boufos*), *Bubo bubo.* Scops Owl, (*Gionis*), *Otus scops.*

16 Vassilikiots occasionally talk about a bird they called *striglopouli* (the screeching bird). 'It is not the owl (*Koukouvagia*)' they told me. Despite persistent efforts I failed to identify the bird's standard name. It is maintained in Vassilikos that 'every time a *striglopouli* sits on the roof of a house and screeches, somebody from that house will die.' Some experienced ornithologists I have met in Zakynthos speculate that *striglopouli* probably is the Barn Owl (*Peploglafka*, *Tyto alba*).

17 The wolf and the fox in those fairy tales take on human characters, and work together, but cheat each other in sharing the spoils. The fox is presented as deceitful and canny, while the wolf is innocent and naive.

18 The traditional system of rights and duties in respect of animal husbandry between a landlord and a labourer (*kopiasti*) – a system practised in the village until the 1960s – included the following obligation: the labourer would be credited with a specific number of animals to 'care' for each season. The landlord would attribute the loss of animals as a result of illness or accident to the labourer's inadequate 'care' for the animals. The labourer would then be expected to replace the value of the lost animals at his own expense.

19 '*Mia mera eidha ena liariko skyli na vazei kato tis katsikes ton Tzaneton*'.

20 '*Enoiosa oti ekana leitourgima giati prostateva ta zoa ton anthropon. Afto to skyli borei na kanei kako.*'

21 Beach marten, (*Kounavi*), *Martes foina.*

22 Turtle Dove, (*Trygoni*), *Streptopelia turtur.*

23 The nine homilies on the *Hexaemeron* were delivered within five successive days in the period of fasting before Easter (Lent*Megali Tesarakosti*). It is customary in the Orthodox Church for *Genesis* to be read during Lent (Sakkos 1973: 16). St Basil's *Hexaemeron* appears to be part of this practice.

24 In the last homily of the *Hexaemeron*, St Basil announces his intention to examine the topic of the creation of man in a future discourse. This task, which was never accomplished by St Basil, was carried out by his brother St Gregory of Nyssa.

25 θ par. 2–4.

26 Plato, Plotinus, Aratus, Theophrastus, Herodotus (Sister Agnes Clare Way 1963: xi). Ailianos,

Diogenis Laertiou, Diodoros Sikeliotou, Opianos, Dioskourides, Philonas o Ioudaios and Hipolytos (Sakkos 1973: 18).

27 See the following works of Aristotle: History of animals, Parts of animals, Movement of animals, Progression of animals, Generation of animals, On plants (in *The complete works of Aristotle*, (ed.) J.Barnes 1984).

28 E par.1.

29 Some authors give the fifth homily of the *Hexaemeron* the title 'The germination of the Earth'. See the 1963 translation by Sister Agnes Clare Way.

30 E par. 2, E par 6, E par. 14.

31 E par. 32.

32 E par. 38.

33 E par. 52.

34 'God also said, 'I give you all plants that bear seed everywhere on earth, and every tree bearing fruit which yields seed: they shall be yours for food. All green plants I give for food to the wild animals, to all the birds of the heaven, and to all reptiles on earth, every living creature' (*The New English Bible* 1970, *Genesis* I 29, 30).'

35 E par. 39–45.

36 Here, the term 'amphibian' is used with its original ancient Greek meaning, denoting a being able to live on both land and water.

37 In contrast with modern taxonomy, internal systems of the animal's body structure are rarely used by St Basil as criteria for ordering animals into categories of related species or genera.

38 Z par. 5–6.

39 The term 'genus' (*genos*) is also used by Aristotle. In his notes to *De Partibus Animalium* I, D. M. Balme explains: 'The root meaning is kinship-group. It is Aristotle's usual word for a type of animal, at every level from infima species to major genus. But he uses it for genus as opposed to species when he requires this distinction... (Balme 1972: 74).'

40 Z par. 7–9.

41 St Basil says: 'The majority of the fishes do not hatch out the young as the birds do, nor do they fix nests or nourish the young with their own labours; but the water, taking up the egg when it has been laid, brings forth the living creature. And the method of perpetuation for each species is invariable and is without mixture with any other nature. There are not such unions as produce mules on land or such as of some birds which debase their species (Z par. 10).'

42 H par. 3–4.

43 H par. 4–5.

44 H par. 12.

45 St Basil demonstrates his point about the respiration of insects with an example borrowed from Aristotle (8.27.605b). He explains that if insects are 'drenched with oil, they perish, since their pores are stopped up; but, if vinegar is immediately poured on them, the passages are opened and life is restored again (H par. 38).

46 Here, the distinction between the 'gregarious birds' and the ones preferring a 'collective form of life' is not made clear by St Basil. Sakkos suggests, after carefully studying the context, that the former category includes those birds living in pairs within large flocks, while the second group includes those birds which live in flocks without a direct correspondence of males and females (in opposite array) (Sakkos 1973: 310).

47 H par. 14–5.

48 See, Aristotle, *The History of Animals* 1,1 (488a) (Balme 1972).

49 H par. 33–4.

50 Θ par. 9.

51 Θ par. 10.

52 Θ par. 8.

53 Θ par. 9.

54 Θ par. 18.

55 Θ par. 18.
56 Θ par. 20.
57 Θ par. 22.
58 Θ par. 23.
59 'The wild beasts are proof of our faith (Θ par. 31).'
60 See 'Animals in Lele religious symbolism' (1957), *Purity and Danger* (1966) by Mary Douglas, and 'Animal categories and verbal abuse' (1964) by Edmund Leach.
61 'The general consistency with which, in nature, morphological differences are correlated with differences in habitat, feeding habits, call-notes, and other aspects of behaviour is the inevitable starting point for any system of animal classification, at the lowest level' (Bulmer in Douglas 1973: 169).
62 Internal body structures, which are an important classificatory criteria for modern taxonomy, have little classificatory importance for St Basil and only in one instance is there a recorded reference to them (see *Hexaemeron* Z par. 5–6).
63 It is worth mentioning here that Aristotle's description of the natural world, despite its naturalistic empirical outlook, is permeated by anthropocentric culturally prescribed hierarchies, similar to those prevalent in *Hexaemeron*.
64 Here, I apply the term 'synthesis' in a deliberate attempt to avoid the problematic use of the term 'syncretism' (see Stewart & Shaw 1994).
65 Brent Berlin has repeatedly argued for the relative importance of perceptual and empirical classificatory criteria, borrowing ethnobiological data from the Aguaruna and Huambisa, the Amazonian communities studied by him and his colleagues (1988, 1992, Berlin & Berlin 1983). But Berlin has not merely confined himself to the exhausting task of demonstrating the universal perceptual foundations of classification. He systematically undermines the relative importance of practical, use-oriented criteria accounted for by ethnobiological classification, creating thus an unfruitful polarity between what he calls 'intellectualist' and 'utilitarian' approaches to classification.
66 Eugene Hunn, although he was among the first to underline the perceptual basis of ethnobiological classification (1976), recognised that 'practically motivated reasoning' was underestimated or taken for granted by anthropologists who overstress the 'intellectualism' of their informants (1982: 830–6). Hunn came to the defence of the 'practical significance' and 'purposiveness' of folk classification and dared to admit that 'pragmatism is no sin' (1982: 830–6); he was subsequently criticised by Berlin (Berlin 1988, Berlin and Berlin 1983) for this position. Morris (1984) was similarly criticised by Berlin (1988) for stressing the 'pragmatic concerns' inherent in the folk biological classifications of the Chewa people of Malawi. The Chewa have a life-form category (*Chirombo*) which accounts for 'useless' beings, a category which would have been perfectly understood and appreciated by my own respondents in Vassilikos. As Morris maintains, 'to understand Chewa folk concepts, one has to accept that they have a pragmatic dimension, and that such taxonomies are not conceptually isolated, as a domain, from other aspects of Chewa culture' (1984: 48). The importance of 'contextual considerations' 'rooted in particular situations' is similarly emphasised by Ellen (1986b, 1993).
67 Under the impact of popular ecology, the Orthodox Church has recently responded with a certain sympathy towards the ecological movement. The Ecumenical Patriarch of Constantinople, 'keeper and proclaimer of the centuries-long spirit of the patristic tradition', published (with the assistance of WWF International) a collection of religious writings entitled *Orthodoxy and the Ecological Crisis*. The Patriarch's pro-environmental position, however, strongly resembles St Basil's discourse of seventeen centuries earlier. According to the Patriarch, 'contemporary man' has 'abused' 'his privileged position in creation', which derives from 'the Creator commanding him to have 'dominion over the earth' (Gen.1,28) (Ecumenical Patriarchate 1990: 1). 'Man', '...the prince of creation' has misused his 'privilege of freedom', and environmental destruction is the result (ibid.: 1). With this publication the Patriarch is making an effort to move closer to the pro-environmentalist position; he is however, inspired by the same principles which shaped St Basil's classificatory account in the *Hexaemeron*.

8
Unlawful hunting

This chapter is concerned with hunting, a little studied topic in the ethnographic literature on Greece, which well deserves to be 'explored in social and cultural terms' (Marvin 2000b: 196). The hunters I examine are Vassilikiot men, whose passionate involvement with hunting, and especially their participation in the turtledove hunt, is renowned in Zakynthos. My objective here is to present a thorough piece of ethnography on hunting and approach hunting as a context of action where men co-operate with each other, compete with outsiders, and negotiate their identities in masculine performances of defiance towards the hunting regulations, the state authorities and the environmentalists.

Besides discussing the men's hunting performances, I will draw some conclusions regarding the contribution of hunting to the rural household and Vassilikiot women's positive attitude towards their husbands' obsessive engagement with shooting wild birds. I will also discuss the discourse developed by the local hunters in response to the accusation that their unrestrained indulgence in hunting is responsible for the decline in numbers of the wild bird population. Needless to say the opposition of the Vassilikiot hunters to the anti-hunting views of the environmentalists is directly related to the presence and activities of environmental groups on Zakynthos and the establishment of the National Park on the island.

I introduce the discussion on hunting in Vassilikos with a poem written by a local hunter about his own hunting experience and the comments I received from other Vassilikiots after reading the poem to them. Then, I present some ethnography on the older generation of Vassilikiot hunters, people who took a unique pride in their sighting skills, their knowledge of game, and in owning their hunting rifles. I continue with a section on the enthusiasm with which today's

Vassilikiot men take part in the prohibited April turtledove hunt and the claims of individual hunters to particular hunting sites. Then, I focus on the Vassilikiot pro-hunting and anti-environmentalist discourse, examining the arguments expressed by the local hunters in support of the idea that hunting is not directly related to decimation of wildlife. In the last section, I attempt to compare hunting with local farmers' attitudes towards the physical environment and the deaths of wild and domesticated animals in particular.

The Vassilikiots discuss a poem about hunting

It was at the end of a hard day's work in the fields when an elderly Vassilikiot unexpectedly recited the following poem to me. It refers to a particular place in Vassilikos, a specific, identifiable tree, with a local Vassilikiot hunter being the protagonist. This same hunter, who is both the poem's author and protagonist, is now dead and the elderly Vassilikiot who recited the poem to me is probably the last person in the village to remember the poem in its entirety. The unusual structure of the poem – some conjunctions, articles and prepositions are omitted – does not represent a particular folk form or literary technique. It is rather the author's intentional invention. By shortening the narrative, he gave an enigmatic, rather humorous flavour, which is characteristic of the Zakynthian satirical but self-critical spirit. The poem relates to several aspects of Vassilikiots' engagement with hunting which I wish to examine in this chapter,[1] although the implicit intention of the poet was to stage and communicate to the local audience his own personal involvement with hunting. In this respect, to paraphrase Herzfeld, 'both the act of [hunting] and the narration that follows it focus on the act itself' (1985a: 16).

After recording the poem, I read it to several other Vassilikiots, men and women I knew well and whose opinion I valued. Most of them had heard the poem before and had related memories to recall. They were particularly pleased with me for recording 'something of their village' which was 'about to be forgotten'. They all agreed that the poem was created because the author wanted to communicate his hunting experiences to his fellow villagers. They all read in the poem the author's desire to stress his own deep engagement with hunting and, simultaneously, to pose a statement about the self as a hunter. They commented upon the hunter's excitement upon meeting a bird, the cuckoo. 'He is like most of us', they said, 'he immediately ran back home to pick up his rifle'. 'Notice how he refers to the characteristics of the gun', they add, 'it was a beautiful gun, bought from an 'English' man'. Compared with the other hunting rifles in the village, 'it was a technologically advanced gun, a new invention' my interlocutors explained, deciphering the cryptic articulation of the poem like literary critics.

All the Vassilikiots to whom I read the poem responded with laughter to the scene in which the over-enthusiastic hunter shoots at the old olive tree. Beyond the comic antithesis – the fall of the small bird, the collapse of the huge olive tree

Πίσω Αγίου Νικολάου Ντόπια κουφάλα
ελιάς, μόσχου παλούκι, μελσσοσυκιά.

Behind St Nikolas's old olive tree, (there is)
the hollow of an olive tree, a stake (made of)
fig-tree wood, a grey bird, a cuckoo.[2]

Πάω σπίτι, παίρνω ντουφέκι διπλό,
μπούκες δύο, βέργα μία, Αγγλος,
ματάκια δύο, νέα εφεύρεση.

I go home, I take the gun, bought from an
English (man), the double-barrelled, two
muzzles, one ramrod, two eyes to see, new
invention.

Πάω πίσω Αγίου Νικολάου ντόπια,
αγκονή, πουλί επανώ... Μπάμ!...

I go behind St Nikolas's old olive tree at the
corner, the bird is above. Bang!

Πέφτει χάμου ο κοκαρέλος, πέφτει
χάμου κι μισή ντόπια.

The cuckoo falls down. Half of the olive tree
falls down as well.

Πάω Αγίου Νικολάου Καλόγερος,
συχώρεση ο Αγιος, δεν το θέλα, το
ντουφέκι το ΄καμε.

I go (to) St Nikolas the monk (i.e. the
monastery), to be forgiven (by) the Saint, I
didn't want to do it, the gun did it!

Πάω εκεί σπίτι, πουλί δεν γνωρίζει.
–Νόνο μην είναι κότσυφας;
–Ανε ξεραΐλα και φάε κιάλλο ένα!
–Νόνο μην είναι πάπουζας;
–Ανε ξεραΐλα και φάε,
το καιρό που ήμουνα μικρό παιδί σαν
κι εσού, με μαθαίνανε οι γονέοι μου
τάξη. Δεν έχεις αράδα να μιλήσεις
μεγάλου.
–Νόνο μην είναι κούκος.
–Ναι, ναι καλά μιλεί μικρό παιδί
κάπου κάπου.

I go back home, no one recognises the bird.
Granddad, that is a blackbird, isn't it?
Hold your tongue and eat (your food).
Granddad, is that a hoopoe?
Hold your tongue and eat ...
When I was a small child like you, my parents
taught me to behave (in an orderly fashion).
You have no right to speak to an adult (per-
son).
Granddad, is that a cuckoo?
Yes, yes, sometimes a small child speaks well.

Ερχεται δεκαπέντε Αυγούστου, κάνει
την ουρά του ... έτσι, κυρίας
Βεντουλέτας!
Μμμ! ο κώλος του παχιός!

Here comes the fifteenth of August, (the bird)
moves its tail (the narrator moves his finger
right and left to demonstrate) (like) a lady
with a fan. Mmm! its bum is so fat!

– lies a statement about the gun's power. The author of the poem wants his audience to notice that his gun was powerful enough to knock down such a huge tree.
Simultaneously, 'he is self-conscious enough to dress up his boasting as humour.
In the following scene the protagonist appears to be a religious man, feeling some
guilt for the damage caused to tree, which is, after all, monastic property. The
hunter displays his guilt by apologising to a monk: 'It was not my fault, the gun
caused the damage'. At this point the Vassilikiots listening to my reading of the
poem will laugh again – not for the hunter's craftiness in dealing with people of
the church – but for his irresistible urge to praise the power of his gun one more
time. Vassilikiots are very receptive to this message, since they have vivid memories of the generation of old hunters who proudly and consistently boasted about
their hunting rifles.

The final act of the poem takes place in the hunter's home. As the brief dia-
logue between the hunter and his grandson suggests, the importance of hunting
in strengthening the relationship between adult men and young boys is immense.

In the poem, the old hunter is persistently interrogated by his grandson about the dead bird. It is taken for granted among the local audience that young boys are interested in hunting. The old hunter further instigates the child's curiosity by denying the young boy's right to talk about the bird. To further stimulate the child's interest, he implies that 'hunting is for men, not for young boys.' The hunter's satisfaction is noticeable when his grandson comes up with the correct answer. In order to reward the boy the old hunter offers further information. He explains that after the fifteenth of August the bird moves its tail in a characteristic way [the hunter demonstrates this by moving his finger], which imitates the ways of an aristocratic lady holding a fan.[3] When the bird moves its tail upwards, he points out, one can see that its rear is fat. The Vassilikiots to whom I read the poem explained that the protagonist here suggests that this is the best time to hunt that particular species of bird. Hunters in Vassilikos become excited whenever they can demonstrate their hunting knowledge. They gain even more satisfaction through 'teaching' their sons, grandsons, or any uninitiated individual, secrets about hunting or ideal hunting spots where game is abundant.

According to the interpretation of the poem by my local audience in Vassilikos, the author's explicit intention was to amuse his fellow villagers and to simultaneously refer to 'things which please every man', such as his special gun, his hunting skills, and his relationship with a grandson. At the same time, Vassilikiots note, that the author of the poem implicitly and humorously invented an occasion to present himself as a knowledgeable and deeply involved hunter. In fact, it was the author's repetitive boasting, rather than the humorous content of the poem that induced laughter and a joyful response from the audience. The Vassilikiots' spontaneous literary analysis helped me most of all in 'sorting out structures of signification' (Geertz 1973: 9) involving multiple layers of meaning attributed to the poet/hunter's engagement with hunting. It also helped me appreciate the double layers of irony posed in the poet-audience interaction – the poet's attempt to hide his irresistible desire to boast with humour and the local audience's humorous enjoyment in uncovering the hunter's boastful attitude.

Old-time hunters and their exploits

It is well known in Zakynthos that three to four hundred years ago, at a time when Vassilikos was scarcely inhabited, the monks of Skopiotissa Monastery (on the local mountain) used to hunt turtledoves and then preserve them in vinegar (*xydhata trygonia*) (cf. Roma 1975). Vassilikos was traditionally described by the town's people as 'the countryside' (*i exokhi*) and many aristocrats would visit it to hunt turtledoves or other game. As the centuries went by, guns became more readily available and hunting became widespread among the poor. The older Vassilikiots, having themselves experienced the remnants of a feudal form of economic dependency upon powerful local landlords, remember how they often had to hide the game they had shot, turtledoves or hares, under their shirts so that 'the master' (*o afentis*) would not notice it. Their 'master' or landlord might have asked them to hand over their prey as a gift for allowing them to remain on his land. Some Vassilikiot hunters maintain that:

> ...in the past, only the rich could afford to go hunting. Poor people had no right to abandon their jobs and join in...so, they would wait for Sundays and other holidays... Some *semproi* (landless tenants) would raise (*sikonan*) the birds for the rich to kill [act as beaters], but they were not allowed to hunt them themselves.

'In the old days', some elderly men go on to explain, 'there were few opportunities for us, the *semproi* (landless tenants), to hunt because we all had a lot of services [to carry out] for our masters.' Thus, the tenant farmers were left with little spare time left for hunting, even when their landlords did not directly prohibit them from indulging in this highly desirable activity.

Despite practical limitations and prohibitions, Zakynthians have always considered hunting a 'passion' (*pathos*) or 'mania' (*mania*), characteristic of their temperament. A Vassilikiot hunter in his early fifties, like several other local men and women who are in agreement with him, emphatically declares that:

> ...everybody hunts on this island. Everybody has a gun in his house; you cannot find a family without a gun! The Zakynthians have a mania for hunting!
>
> When I was a child, I used to wander in the fields with my sling, shooting whatever I could find. We used to hunt turtledoves, mistle thrushes, woodcocks, hares or even robins.[4] We used bird limes for robins and other tiny birds (*lianopoula*). We had snares (*vrohia*) for turtledoves made of hair from a horse's tale. The turtledoves, tired (*kourasmena*) from their long journey, used to drop into the snares which were placed on the trees, anywhere where there was space for the birds to perch.

Using snares, lime-twigs and other kinds of traps, Vassilikiots successfully hunted turtledoves, small birds and hares (cf. Kenna 2001: 33). Numerous younger and older men described to me their skill in improvising and inventing new kinds of traps, using wood, leaves and stone, in accordance with the requirements of hunting particular game in particular places. Since guns and ammunition were scarce and expensive in the past, traps were an alternative means of catching wild animals or birds, a valuable source of meat for poverty-stricken families. Most domestic animals entrusted to the 'care' of landless tenants were the property of the landlord, and had to be 'kept' alive either as working animals or as capital to be maximised. Consequently, the trapped prey was valued by the rural household as a supplementary subsistence source. Even tiny birds, like the robin, when caught in sufficient numbers,[5] would provide any given family of farmers with an extra meal. As one elderly woman puts it: 'Had people in the old days not been crafty, they would not have made it. They had no money for shot, but they caught a lot of birds with traps.'[6]

Traps were usually set by young boys who were eager and impatient to exercise their hunting 'passion', but were unable to buy a gun. Adult men were also interested in traps; for instance, the snares for turtledoves were mostly set by adults. I was surprised to find out that these traps were, in fact, highly effective techniques for capturing wild birds. In Vassilikos and Keri (another Zakynthian community) great numbers of turtledoves were caught in the past using snares. Nowadays, this type of hunting is prohibited by state legislation and is abandoned in both places. Most Vassilikiots appear in general uninterested in setting traps, although they still enjoy narrating the 'trapping exploits' of their youth.

Hunting rifles in the old days, like traps, required a lot of preparation and *meraki*, a word that could be roughly translated into English as artistry or good taste (cf. Sutton 1998: 73). Here, a couple of Vassilikiot hunters elaborate on this, arguing that a great deal of time had to be spent on the preparation of ammunition:

> At that time we had single-barrelled rifles (*monokana opla*). We had to load shot with gun-powder and pellets. There was a special instrument used for this job. Those guns were dangerous; you could lose an eye, or a finger if they went off.
>
> I have been hunting since I was a child; I was using muzzle-loading guns then (*kynigousa apo paidhi me ta emprosthogemi*). There were few bullets at that time. One had to sit down and make the ammunition oneself.

At that time, most landless tenants in Vassilikos would have regarded hunting rifles as their most valuable possession. They were objects of display, signifying one's hunting skill and involvement in hunting. During my fieldwork I often heard Vassilikiots commenting upon the importance of hunting rifles for the 'old-time hunters' with statement such as this:

> Those people used to carry their guns to the coffee shops, and hold them on their knees or place them upright by their side. They used to bet on their ability to aim at various targets (*sto simadhi*). Those old-timers (*oi palioi*) were terrific (*tromeroi*) hunters!

Carrying a gun, especially a unique one, was a statement about the self as a hunter and one's ability to hunt. The owner of the gun should, ideally, be prepared to demonstrate his shooting skills whenever challenged by others. Lefteris, my adoptive father in the field, once told me about the following incident, which took place only a few years ago:

> I was hunting down at Longos [a wood]. A man from the town approached me. He was riding a motorbike. He noticed my gun and challenged me: 'Why are you carry-

ing this gun, since you are no marksman (*afou dhen xereis simadhi*)!' I told him: 'Throw your key-chain with the pen-knife in the air and, if I miss, I will give you a hundred *drachmas*.' Adas, another local man, was around with his sheep and said to the man from the town: 'Take the key of your motorbike off the chain. Otherwise you will not be able to get back to your home.' The man from the town was hesitant (*dhistakhtikos*). Adas insisted and eventually the man from the town took his key off the chain. I hit the chain with the core of my shot (*smparo*) and nobody saw the chain again. It was thrown up with force towards the wood. Then this man said to me: 'Do you really want me to give you a hundred *drachmas*? Do you know how much the chain and the pen-knife cost?'

All the Vassilikiots I know talk about the old-time hunters with awe. They comment upon the intrepidness (*palikaria*) of those men and their hunting skills with admiration. In the context of all male gatherings, hunting skill is acknowledged to be a source of respect, an integral part of a man's socially defined identity. Stories such as the following, indicative of some men's involvement in hunting, are frequently recalled in local conversations. A senior hunter remembers:

My father and his younger brother, Barmpa-Giannis, were both great hunters. Barmpa-Giannis though, was the best hunter in the village. He could shoot a chick-pea or a *mirtokouki* [another seed] out of the air. Other men used to bet on his skill.

One day both brothers were sitting in Shourpou's shop [a coffee-house] with their guns on their knees (*sto gona*). A quail came and sat on a fence nearby.[7] The two brothers started arguing about who would shoot the bird. Everybody in the coffeehouse argued that my father should have a go since he was the older brother. Barmpa-Giannis agreed saying: 'But be careful not to miss it.' My father shot at the quail but he missed. Barmpa-Giannis didn't speak to him for a year!

Contemporary hunters

Unlike the past, when the villagers were constrained by poverty and feudalism, the present-day Vassilikiots have plenty of time to devote to hunting. In fact, they arrange their agricultural activities so as to secure enough free time to participate in turtledove hunting, the most important seasonal hunt on the island. Nowadays, the hunters no longer spend time preparing shot or setting traps; instead, they take advantage of the availability of an abundance of technologically advanced guns and related equipment. Thus one can legitimately claim that present day hunting involves more action and less preparation. But still, as in the past, and this is the most important point, hunting is considered to be a central feature of men's life in Vassilikos.

The favourite discussion in the coffee-houses, where men gather in the late afternoon after work, is about hunting. It is more popular than politics, for a political discussion is bound to cause a quarrel, whereas the hunting discourse has a unifying effect. Discussions focus on subjects such as the number of birds killed at particular hunting spots, hunting rifles and dogs, or specific kinds of game. Real events that occurred during hunting are described in detail. The protagonists of these narratives are local people who display their hunting skill or other traits of their personality in their narration. Vassilikiots, who enjoy discussing people they know, take any available opportunity to joke and tease each other with pointed comments about success or failure in hunting. In this social context, the hunting skill of each hunter is constantly assessed and reassessed, while individual hunting experiences gradually become shared properties in the community.

Turtledove hunting is a major issue in Vassilikos. Most men look forward to the two seasons of this hunt. The first is in April and the second from mid-August to the end of September. Given that the numbers of turtledoves had been decreasing in recent years, the state authorities and the Zakynthian Hunter's Society have come to a mutual agreement to ban the April hunt. In practice however, despite the severe prohibitions, turtledove hunting has not been restricted in any way. Some Zakynthian hunters are brave enough to walk past the Prefect's headquarters in the island's capital with their guns to demonstrate their refusal to comply with the laws banning hunting. 'Although there are not many turtledoves left' the hunters admit, 'we will go on hunting, because this is an important part of our life.' 'Nobody will ever dare to stop us', they declare, while enjoying the relative security of male solidarity in the coffee-house.

Vassilikos is one of the most important hunting sites on the island. It is the first meeting place for turtledoves on their migration route over Zakynthos. Every year, some days before the April hunt, an air of excitement spreads all over the village. One can feel that something important is about to happen. Soon comes the day when the local hunters take their positions in their hunting posts, armed with guns. Turtledove hunting has started. Along the main road of the village, in the olive groves, deep in woods (*logoi*) and higher up on the rocky hills, hunters can be seen waiting patiently with their guns for the long awaited turtledoves to appear. On those rare occasions when the forestry department's patrol car approaches the village, the hunters, whose presence was previously conspicuously manifested, disappear. Every car that heads towards Vassilikos on the single village road can be seen from the neighbouring houses and the message is easily spread by telephone or other means.[8]

The house I was living in during my fieldwork was situated in an olive grove right at the centre of the turtledove hunting field. As a result, I had the privilege of experiencing the turtledove hunt at Vassilikos 'at close quarters' for two consecutive seasons. Here I will present some extracts from my fieldnotes:

> Tired of the repetitive noise of hunting rifles I was on my way for a walk in the fields nearby. A Vassilikiot hunter, locally renowned for his masculine performances in a variety of contexts, was positioned on a roofless wooden platform at the top of an olive tree. Covered with leaves, and dressed in a camouflage uniform, the hunter spent the

whole day shooting at passing turtledoves and the spent pellets from his misses were falling on the roof of my house. He tried to appear talkative – a serious compromise of his reticent style – to measure my reactions. He started talking about a documentary he saw on the television about 'those black people in Africa'. Proud of himself for watching a documentary [of an educational character], he appeared eager to share it with me. As an educated man I was expected 'to know about those things'. He described to me – what else! – scenes of hunting in Africa. He talked with admiration about a huge black hunter who killed lions; the African hunter was tall and muscular (*me kati myes na!*) and he waved his own impressive muscles at me to illustrate his point. He also commented on the poverty of those African people and 'the conditions under which they live'. He ended – and that was his intended message – by stating the popular local scenario in which African people are 'destroying the turtledoves'. 'People in Africa poison the turtledoves to safeguard their cultivation. They are poor people who are starving. This is why there are so few turtledoves left!', the hunter concluded. He appeared apologetic for hunting turtledoves as there are so few left, although I didn't try to make him feel guilty about this fact. At the same time he was expressing his anger for having so few turtledoves to shoot at!

For many subsequent days hunters continued to shoot over my house. The deafening sound of the rifles was matched by the rhythmical rolling of the pellets over the roof. There were moments where 'I felt that I had gotten a taste of life in a war zone', to use the expression of David Sutton who had a remarkably similar experience during his study of the Easter dynamite throwing in Kalymnos (1998: 57). Most of the shooting originated in the neighbouring olive grove, which had been hired by a group of hunters from the town. They constructed a primitive shelter made of tree branches and leaves. They waited in their shelter for turtledoves to approach. I could hear their conversation and jokes. I wrote in my fieldnotes:

Although today is Easter Day, the most important religious celebration in this country, hunting still goes on. I am surprised by the fact that so many men leave their families – women, children and old men celebrating at home – in order to come hunting for the whole day. A group of hunters is shooting thirty or forty metres away from my front door and the noise is particularly annoying. Some of my relatives and I are hiding indoors, afraid of gunshots coming from every possible direction. We can even hear the sound of pellets falling on the roof and in the garden. The hunters appear to me to be intoxicated with a distinctive Bacchic fervour. I am able to guess the time each group of turtledoves approach the area by the sound of guns shooting from various distances and various other noises. They shout out when they shoot: 'I've got one' (*to faga to na*), and I can hear a second voice replying, 'I've got one as well, four or five of them have just gone over' (*ki ego efaga ena, perasan tessera-pente*). Every time a hunter fails to kill a passing turtledove, he warns his companions that the bird is approaching; 'One is coming over to you' (*sou erkhete ena*) one hunter communicates to the other. I admire the cooperation between local men during hunting. Yet, my degree of empathy has by now been exhausted.

On the same Easter afternoon, I was sitting in the front yard of my house with some relatives, when a hunter from the town attempted to shoot a turtledove that was flying over our heads. Angry with him for aiming his gun towards us, I dared to complain. A local hunter, the same individual with the theory about Africans exterminating turtledoves, who was equally annoying me with his shooting, came to my defence. In his deep masculine voice he yelled at the hunters from the town: 'You shoot at a house? People live in it' (*to spiti varate? anthropoi zoun mesa*). That same afternoon the local hunter confessed to me that he had been looking for an excuse to vent his anger at the hunters from the town. Unable to hide his antagonism towards the latter he confided to me: 'They are not real hunters like us. They just have fancy guns and money to hire good hunting spots, but they don't know much about this hunt.'

The hunter from the town, who aimed his gun in my direction, had hired a piece of land to use as a hunting spot from Dionysis (a pseudonym), a key informant of mine. The land was the property of a landlord of noble origin, but Dionysis, being the landlord's tenant (*sempros*), was responsible for its cultivation (*sempria*). Although Dionysis was a valuable informant and I was reluctant to endanger our friendly relationship, I broached my complaint about this particular hunter from the town. Dionysis had already been informed by the hunter and had prepared his argument beforehand. He said that the man whose house I had rented also went hunting. 'Besides, your landlord can have no control (*dhen borei na kanei koumanto*) over the neighbouring property just because he has built houses so close to it', Dionysis argued. From this I could tell that he had already had a discussion with my landlord, who had complained on my behalf. Dionysis tried to reassure me, in the presence of other men in the coffee-house, that 'the hunters were aiming at the birds, not at people.' He explained to me that falling pellets were not dangerous since they had lost their force. However, Dionysis refused to accept my complaints about the noise of the guns: 'Complaining about the noise is too much' he argued decisively.

As this short story suggests, the competition among hunters for securing suitable hunting positions during the turtledove hunt is subject to local rules of conduct and respect. The local hunters exercise their prior right over the allocation of the most desirable 'hunting posts'. They retain hunting posts on their land, in cultivated fields that they have secured through tenancy agreements, and in a local wood that belongs to a foreign corporation. In the first case, they feel sufficiently confident in their claim to the hunting posts to rent these hunting posts to outsiders or invite friends to hunt; especially those friends to whom a favour or obligation is owed. I remember a friend in Vassilikos commenting about a fellow villager:

> Look how many people hunt at your landlord's place; they are all friends of his from the town. They shoot all day and cause trouble (*bela*) to him and his wife. But what can the poor man do about this? He owns a restaurant, as you know... you understand, he cannot turn away the friends who impose themselves on him (*pou tou fortonontai*).

Rights over hunting posts in the local wood (*logos*) are established by the active presence of the local hunters in the area and the frequent use of the hunting posts by them. This land was bought by the Tourist Corporation a couple of decades ago with the aim of developing it for tourism. However, this development never took place owing to doubts about the legal status of the transaction. The landlord who previously owned the wood is now claiming it back. In the meantime, local men settle their own 'hunting claims' on this disputed terrain by use of their own local code of 'hunting conduct'. A young Vassilikiot hunter who hunts frequently in the local wood elaborates:

> '*I Maliari Petra, O Paliolinos, Ta Xera, I Omprela*' [placenames] are hunting posts (*posta*). These are only a few; there are many more as you probably know...
>
> You can never take the hunting post of a local man. A local man, however, can rent his hunting post to strangers (xenous). He can do that if he feels confident (*sigouros*) of his hunting post. You secure your hunting post by going there frequently.
>
> The non-local hunters always have to respect the local ones. They have to adjust to their rules. They can't do otherwise!

The use of the local wood as hunting terrain, like a few other mountainous parts of Vassilikiot land, is not strictly controlled by property titles or tenancy agreements. Yet, the active presence of Vassilikiot hunters is significant enough to establish claims over particular hunting spots. As the young hunter quoted above has already explained, those claims include the right to rent the hunting spots to outsiders. Considering the fact that the April turtledove hunt is officially an illegal activity, the legitimisation of the local hunting *status quo* and the overconfidence with which the local hunters control their hunting resources, are conspicuous examples of the celebrated local defiance of the law and of the State's authority. It is also an implicit attempt to undermine the efficacy of environmental conservation and, as I shall demonstrate in the following section, an explicit challenge to the presence (and anti-hunting stance) of the advocates of environmentalism on the island.

Hunting despite prohibitions

During the period of the turtledove hunt the dominant topic of conversation in Vassilikos is about – what else? – turtledove hunting. The major concern of the hunters is the reduction in the number of birds in recent years, which is threatening to deprive them of their favourite 'passion' and pastime. An elderly shepherd, who is also a hunter, was eager to comment on the situation, while pasturing his sheep in the fields of Vassilikos:

> The turtledoves are few, the guns are many. I took my gun with me twice while I was out with the sheep but then I left it behind. At noon a few turtledoves arrived,

exhausted by the heat. They shot them at once! In the past the olive groves were full of them.[9]

A younger hunter reflected on the same topic:

> There are few turtledoves left. Hunting must take place only in August... *but* because there is no other important game on the island the authorities are tolerant.[10] You cannot take hunting away from the people... it is such an important part of their lives.

Most hunters in Vassilikos do not consider hunting as directly responsible for the reduction of the turtledove population. They propose an alternative explanation, which I recorded in my fieldnotes as the 'pesticide rhetoric'. The argument goes that pesticides are to be blamed for the decline in game; and this is indeed a popular argument among hunters in modern Greece. In Vassilikos, people are conscious that pesticides and other chemicals can have devastating consequences on the local fauna. A couple of decades ago, technical advice on pesticides was inefficient and their introduction in Vassilikos was accompanied by mistakes in their management. A farmer, who is also a renowned hunter, illustrates:

> The big landlord, instead of ploughing the land, threw 'poison' [*farmaki*: he means pesticide] on it to get rid of weeds. He found all the birds and insects dead on the ground. He regretted it and he didn't do it again. Another year he made a similar mistake. He put more 'medicine' [*farmako*: i.e. pesticide] for *dhakos* [a disease affecting olive trees) on the olive groves. All those birds which came and sat in the trees died. He is now careful about giving the correct dosages (*dhosologia*).

Most Vassilikiot hunters, in fact almost all hunters, blame pesticides to one degree or another for the decreasing numbers of wild birds. If they are asked about the relevance of unconstrained hunting to this issue they repeatedly state their conviction that hunting is not, and could not be, the major cause of the decline in wild fauna. Here is a typical response:

> There are a lot of guns in the village, more than any other time...but there used to be many birds as well...
>
> There were many birds in the past. The turtledoves were like a cloud over the olive groves. Now, can you see any? People have always hunted on this island, but the birds were always plentiful.

'Look at the sparrows', other hunters argue, 'think of how many they used to be in the past! Why are there so few of them now?' After a small pause that adds greater force to their rhetorical question, they conclude: 'The sparrows were mowed down by the chemicals [pesticides].'[11] The sparrows are not considered as game in Vassilikos. Thus, their decline in numbers is treated as evidence that hunting is relatively irrelevant to wild bird population densities.

Arguments that blame pesticides for the decline in turtledoves rely on some empirical evidence. No one can deny the fact that pesticides eradicate many species of insects, which provide the wild birds with food. Moreover, some pesticides directly poison grain-eating birds. The pesticide rhetoric, however, acquires its real significance when it is placed in the context of the widespread conflict between hunters and conservationists in Greece. The 'ecologists', to use the local generic term for the conservationists, advocate a novel set of moral categories concerning hunting and the protection of wild animal and birds. An ecologically oriented 'ethos' is championed by the media, where the protection of fauna and flora, 'the national natural heritage', is frequently and systematically promoted. This new eco-friendly discourse is well received by the urban public, while, at the same time, the ethics and morality of environmental protection is openly encouraged in schools and educational establishments throughout the country.

The hunters of Vassilikos have reason to feel threatened by the rise of the ecological discourse, not because the 'ecologists' have the power to restrict hunting in practice, but because hunting is deprived of its positive moral connotations. A cultural practice traditionally considered as positive, is now treated by a growing number of urban neighbours as undesirable, destructive behaviour with a negative moral stigma attached to it. It is not surprising then that the hunters, needled by the environmentalist discourse, react by formulating their own alternative discourse, which is negotiated and communicated at both the national and the local level.

The more widespread, nationally defended discourse in support of hunting is championed by educated, and primarily urban-based, hunters who ideologically promote the practice of hunting. Their arguments are derived from popular beliefs, historical sources, or even 'ecological' studies and statistics, creatively reinterpreted, or even, in some cases, misinterpreted. Educated urban hunters publish their views in newspapers and specialist magazines[12] that contain well-organised and detailed information about guns, hunting dogs, the biology and habitat of wild game and personal stories reflecting the engagement of particular hunters with various kinds of game. In addition, these magazines provide a forum for hunters to express their dissatisfaction with, or their counter arguments against, the views of environmentalists. The latter are biologists, conservationists and eco-activists, who frequently make their allegations against hunting in newspapers, popular ecology magazines and on television.

More systematic attempts to establish a coherent pro-hunting discourse (Kampolis 1991, editorials in the magazines mentioned above[13]) employ a selective variety of data from anthropology, history, or even psychology, in a systematic attempt to argue the importance of hunting as an indispensable part of human life. The fervent and politicised nature of the pro-hunting arguments parallels the moralising discourse of the environmentalists. In fact, the former is instigated in direct response to the latter. Pro-hunting articles in newspapers and specialist magazines typically start with a reference to particular allegations made by the 'ecologists' (*oikologoi*), the 'pro-ecology-advocates' (*oikologountes*), or even renowned popular-ecology theorists. The arguments aim to demonstrate that hunting and hunters are not responsible for the decline in Greek fauna – as the 'ecologists' 'unjustly' claim – but that this is due to other factors, such as pollution, industry, pesticides, or even, unwise measures taken by the 'ecologists' themselves. The published arguments in support of hunting attempt to confront rationally the statements of the environmentalists and simultaneously strengthen and reinforce the practice and ideology of hunting among the hunters themselves.[14]

The Vassilikiot discourse in favour of hunting, although evidently less structured and systematic, nevertheless follows a similar course to the more widespread one already described. The emphasis is again on the ideological defence of hunting against those – the environmentalists or 'ecologists' – who attempt to undermine its moral foundation. Vassilikiot hunters attempt to spread the responsibility for the decline in the wild fauna by emphasising other factors causing environmental degradation. In particular, pesticides and foreigners are most often accused of being responsible for the reduction in the numbers of wild birds. The scenario involving Africans who 'destroy' the turtledoves, articulated by one hunter in the previous section, is in fact a very popular explanation of the turtledoves' decline among Vassilikiot hunters. The pesticide rhetoric, locally referred to as 'poison' (*farmaki*) or 'chemicals' (*khimika*), is another popular way of accounting for the drastic reduction in numbers of all kinds of game over the last decade. Any observable correlation between the increase in hunting guns and the decrease in wild birds, although taken into account in the local discourse, is treated by Vassilikiot hunters as fortuitous.

Arguments emphasising the importance of hunting for Vassilikiots' life are also employed. They are expressed, however, in a very rudimentary form. 'We always used to hunt' or 'you cannot take this [hunting] from us' is what the local hunters claim, practically unaware of the power this kind of cultural valorisation can have. They prefer to blame the environmentalists instead, for maintaining their own culturally specific morality and assumptions. I conclude this section with the words of a local hunter. His references to the farmers' 'care' (*frontidha*) for their own domestic animals provides the ideal bridge linking the present discussion with the following section. Here the local hunter is unravelling – in terms of a discourse replete with 'naturalisations' (see, Yanagisako & Delaney 1995) – the indigenous philosophy concerning hunting, 'ecology' and the principles of caring for amimals in the environment of the farm:

> I regret the bird I kill (*to klaio to poulaki pou to skotono*). I regret every bird I kill but this is how life is. Look at this chicken [he points with his hand at a chicken roaming around his yard]. It will die in eight months.
>   I care for this chicken. I feed it, I provide water for it (*to potizo*). In eight months it will die. This is its nature (*i fysi tou*). It has a life, a good life. I provided everything for it.

It has a good life. A natural (*fysiki*) life. It grew up and lived. And then it is the time to die. Where else should the chicken go?

Chickens reproduce. This is why they make so many chicks. There is no place for more. It is natural (*fysiko*) for them to die.

I raised the chicken, it gives life to me now. It is the same with the turtledoves. But the African people (negroes: *arapadhes*) poison (*farmakonoun*) millions of them.

What will ecologists do about this? The ecologists do not deal with the threats (*kindhynous*) to nature. The ecologists are only concerned with their pockets.

Look at this beauty around you [he points at the cultivated land and olive groves]. This is ecology. Who looks after the maintenance of this...'

Hunting and farming

In Chapter Six I examined the importance of the notions of 'care' (*frontidha*) and 'order' (*taxi*) as concepts relevant to the animal-human relationship in Vassilikos. I also demonstrated how the Vassilikiot farmers do not make a clear-cut distinction between care and labour spent on their 'own' animals and any sentiments of affection towards them. Animals incorporated into the context of 'care and order' established by the farmers – even animals which were once wild – are entitled to the farmers' protection as members of the rural household, and bound to it with ties of 'reciprocal obligation', to quote du Boulay (1974). However, animals which exist outside the context of 'care and order', are treated in most cases with detachment or hostility.

Unlike harmful or non-useful animals, however, wild birds and hares comprise the only available game in Zakynthos and are evaluated by Vassilikiot farmers in positive terms. 'These are useful animals', Vassilikiots say, 'they are edible (*trogontai*).' Thus, edible wild animals, like domestic animals, are treated as 'useful' (*khrisima*), since their eventual death can contribute to the welfare of the rural household. But edible wild animals, unlike domestic animals, exist independently of the context of 'care and order' established by the farmers, and consequently the farmers feel free to appropriate their 'usefulness' whenever they are in position to shoot them and without being constrained by any previously defined plan or orderly arrangement.

My observations suggest that quite often the actual process of killing and consuming hares and wild birds does not differ significantly from the process of killing and consuming free-range rabbits and poultry. In both cases, men are expected to kill the animals and women to prepare the killed animals for eventual consumption. Similarly, in both cases, friends or relatives of the household, or people to whom the household owes an obligation (*ypokhreosi*) are invited and the meat consumed is valued as being 'special' (*xekhoristo*). Where domestic animals are killed, the farmers communicate to their guests their pride at being in a position to consume food produced on their 'own' farm. Where hunted game is consumed, the farmers praise the quality of the 'wild animals' meat and the skill of the individual hunter.

The following ethnographic examples will illustrate the similarities between killing and consuming wild birds and domestic free-range chickens:

> Lefteris, my adoptive father in the field, was about to kill a chicken. His daughter-in-law was in a hurry. They were expecting guests from the town and she had to do the plucking because Lefteris's wife was absent. Lefteris took his gun and asked me if I wanted to join him.
>
> 'Why are you carrying the gun', I asked him, 'I thought we were about to kill one of the farm chickens'.
>
> 'They can't be caught during the day. Try if you want!', Lefteris replied.
>
> We walked around his farmland at a slow, purposeful pace. Lefteris was trying to find an appropriate chicken to shoot but this task was not easy. The chickens were hiding at the sight us, aware that we were after them. I felt we were out for a real hunt. There was a strong feeling of expectation in the air.
>
> At last, he found a suitable chicken. 'Silence', Lefteris told me and with the agility of a young man he shot it. I gathered up the dead chicken and I took it to his daughter-in-law to pluck it in hot water. She expressed her satisfaction with the particular chicken 'because it was big and young' and started the cooking preparations.

A few months later I participated in a similar event. I followed Lefteris on his way to hunt his 'own property' (his chickens) on his 'own property' (his farmland). This time he was accompanied by his hunting dog Moros. He killed two chickens with the same shot in a way that resembles killing several wild birds with one shot. He said 'with one shot, two turtledoves' which is a common Greek proverb about realising a double objective with a single effort.[15] One of the dead chickens fell on a bench and Lefteris commanded his dog to collect it and the dog retrieved the bird successfully. This particular chicken hunt was very similar to a proper hunting expedition.

In the previous chapter, I described an incident in which the same protagonist had caught 'with his bare hands' a hare hiding on his land. He was extremely experienced at catching free-range domestic rabbits roaming on his farm. He carried the hare around the farm, holding it by the ears in the same way he carried his rabbits. Then, Lefteris announced that if the hare was female [male hares are expected to behave antagonistically towards male rabbits] he would 'keep' it alive and let it mate with his tame rabbits. But since the hare proved to be male, he was persuaded by his wife to kill it for immediate consumption. Lefteris, his family and I enjoyed the hare's meat with friends the next day, while everybody praised Lefteris for his skill in caching a wild animal with his bare hands.

As these examples illustrate, hunting and killing domestic animals on the farm cannot be seriously differentiated in the context of daily life. Hunting often takes place on the farm and several Vassilikiot men carry their guns around while executing their various daily farming tasks. Vassilikiot women welcome the game birds in the home in the same way they accept killed poultry: they pluck them and plan 'how to cook them' and 'whom to invite' to the prospective meal. As the following description demonstrates, wild birds and domestic chickens are often consumed within the forty-eight hours, both kinds of meat being considered fitting meals for a special occasion. Here is another example:

> Dionysis had just arrived home. He was returning from hunting. He patiently stayed on Mount Skopos all morning waiting to shoot any birds. He managed to bring back about ten birds: a few blackbirds (*kotsyfia*) but mostly thrushes (*tsikhles*).
> 'Thrushes stay in Zakynthos from November to March' he told me, 'they are very tasty! Why don't you come tomorrow to eat with us at noon?'
> He gave the wild birds to his wife to pluck them and prepare them for tomorrow's meal. His wife offered him one of 'their own chickens' cooked in the oven with potatoes. She said: 'We kill chickens on our farm, quite often; it is good that the chickens we eat are our own chickens. We killed this one to celebrate our son's name-day. Tomorrow we will cook the thrushes for you...'

Having emphasised the apparent similarities between hunting hares or wild birds on the land around Vassilikos and killing poultry or rabbits on the farm, I now wish to clarify some of the fundamental differences between those two sets of activities. The difference is rooted in the importance of the context of 'care and order' to which the domestic animals are introduced. Vassilikiot farmers are highly selective about which farm rabbit or chicken to kill. Before arriving at such a decision they consider the gender, age, stage in the reproductive cycle, behavioural traits and appearance of the animals in question. Often the animals to be killed are identified well in advance, being in most of the cases young male rabbits or cocks, or even old female animals/birds that have already fulfilled their reproductive potential. In other words, decisions concerning which animal to kill are dependent upon criteria that prioritise the benefit of the farm and are in accordance with the ideal of the household's self-sufficiency. According to the farmer's understanding, the domestic animals destined to die have already received from their human caretakers the appropriate or expected amount of 'care' and are now in a position to reciprocate this care with their orderly planned death. As the hunter quoted in the previous section vividly explained: 'I raised the chicken in the first place, it gives life to me now!'

Killing wild birds, however, is an activity independent of the constraints of 'care and order' on the farm. Hunting those birds fits perfectly well with criteria that relate to the benefit of the farm and the ideal of the household self-sufficiency. But unlike the case of domestic animals, the availability of wild birds is determined by their natural seasonal migration, rather than by the timing and 'order' established by the farmers. As parts of a conceptual domain that exists independently of the human-made 'order', wild birds, like storms, drought, weeds and unconstrained vegetation are legitimately approached by Vassilikiot farmers and hunters with a confrontational attitude, reminiscent of their daily 'struggle' (*agona*) to bring order to the uncontrollable aspects of their immediate physical environment (see, also, Chapter Four).

Conclusions

Hunting is a regular annual activity in Vassilikos, one considered by the local inhabitants, and especially the men, to be of special importance. It has been practised in the area since the times of the Venetians (1485–1797), but then it was largely the prerogative of the rich and powerful, who had the time and means to enjoy it. In this century, however, hunting has become popular among the poor, although, until twenty or thirty years ago, rifles were rare and valuable possessions, available only to the most committed and esteemed hunters in the community. When my older informants were young, an older generation of Vassilikiot hunters were the eminent protagonists of the local narratives and poems, recited at family gatherings or in the little all-purpose premises which served as coffee shops. Old-time hunters were respected for their shooting skill and renowned for their 'boasting' (*kafkhisies*) over their guns and hunting achievements. Their boasting was humorously tolerated by their wives and their male friends, and their skill was admired by young boys and younger hunters.

Since then, hunting in Zakynthos has been drastically transformed from an aristocratic pastime into a celebrated 'passion' (*pathos*) shared by the vast majority of the male population – 'hunting passion' being an equivalent to the expression 'hunting fever' used by North European hunters (Hell 1996: 209). Guns have multiplied, game decreased (cf. Kenna 2001: 149), and hunters have become more emphatic about their commitment to and involvement in hunting. Confident and proud of their engagement with hunting, present day Vassilikiot hunters arrange their farming or tourist business so as to secure the time required for the pursuit of their hunting objectives. Neither their wives, who do not appear particularly threatened by their husbands' time-consuming engagement with hunting, nor the legislation which aims at curtailing hunting activity, have diminished the local passion for hunting. During the turtledove hunting season, especially during the prohibited April hunt, Vassilikiot hunters make their presence felt with their guns and their uncompromising postures.

Hunting is a celebrated male endeavour in Vassilikos, an activity undertaken solely by men (cf. Handman 1987: 220; Moore 1994: 83).[16/17] It can be accurately described as comprising a context of action including and concerning men, an all-male sphere of activity, not very dissimilar from the coffee-house as this has been studied by Papataxiarchis (1988, 1991). In this author's writing, the coffee-house is distinguished as an egalitarian, almost anti-structural social context, in which masculine identities are shaped and reinforced. While relationships and alliances between clusters of related women in the matrifocal neighbourhoods of

Lesbos are governed by kinship, Papataxiarchis (1991, 1995) explains, male solidarity in the coffee-house is ruled by friendship and commensal equality. The egalitarian character and the masculine-identity-formation potential of commensal all-male gatherings is further recognised by Gefou-Madianou (1992a:11–2), while Loizos (1994:77), similarly comments upon the coffee-house's less structured and less hierarchical constitution when compared to hegemonic institutions, such as the church and the state.

Like the coffee house, hunting, as I have encountered it in Vassilikos, can accurately be described as a specific context among others – to follow the approach offered by Cornwell & Lindisfarne (1994) and Loizos (1994) – where male identities are asserted and reinforced.[18] Herzfeld, in *The Poetics of Manhood* (1985a), underlines the importance of the performative aspect of 'being a man'. According to this perspective, the theft of an animal carried out by a Cretan villager and its subsequent narration, aims at a demonstration of the quality of the act itself and the skill of the protagonist (ibid: 16–8). Likewise, hunting in Vassilikos can be understood as a culturally defined stage upon which personal identities are negotiated, and the stage itself, the performance of hunting, as an 'agonistic spectacle' (Faubion 1993: 220). It provides opportunities for the Vassilikiot hunters to articulate their claims to manhood (cf. Moore 1994: 84), to affirm their friendship with other local men and compete with outsiders, to perform and to recount their achievements. Boys are given a chance to fail, to try again and to succeed. Adult men can seize the occasion and excel by becoming more successful men. Their masculine endeavours as portrayed in hunting will eventually become part of the local history, repetitively celebrated in local narratives, or even, poems.

During the turtledove hunt, small groups of men hunting together, joking and enjoying the all-male company are visible to any observer in the fields of Vassilikos. When hunters miss shooting a passing bird they warn their comrades, who are waiting in neighbouring hunting posts, of the imminent approach of the bird. At the end of a successful hunt, a group of hunters may retire to the house of one of the hunters, where part of the game is jointly consumed and the hunting achievements of the day are recounted. Hunting unites rather than divides the local protagonists, who celebrate male solidarity much like men in the commensal atmosphere of the coffee house. As with Sofka Zinovieff's informants, who hunt foreign women rather than birds, Vassilikiot hunters can be described as enjoying 'the planning, the discussions, and the competitive equality that form the base of the activity' (1991: 206).

Cases where competition over hunting skills can lead to a serious quarrel between two fellow villagers are rare. The incident of the two brothers who did not speak to each other for a year, because one of them missed when shooting a quail, is remembered by present day Vassilikiots as a rare example, an exaggerated aberration. And exaggeration, in most of the hunting narratives, is well received by the local audience. Vassilikiots, both men and women, enjoy listening to the 'boasting' of local hunters and the exaggerated hunting stories. Narrators themselves – like the poet-hunter whose poem was discussed earlier in this chapter – employ their mastery in exaggeration in a self-critical humorous manner, conscious of its performative character. As Garry Marvin has noticed 'there is an aesthetic and expressive quality and a dramatic structure to hunting that elevates it beyond the utilitarian and the mundane' (2000a: 108).

To my initial surprise, Vassilikiot women did not appear to complain about the obsessively prolonged involvement of their husbands with hunting. On the contrary they clearly appreciate the contributions hunting makes to the household economy. They receive the game from their husbands with pleasure and proceed to make plans about cooking and the guests to be invited to the meal. Unlike all-male gatherings at the coffee-house and other examples of Greek hunters who exclude women from the consumption of game (Handman 1987: 220–1), Vassilikiot men – who primarily hunt small birds or hares in the vicinity of their households – most often choose to celebrate their success in hunting in the company of their wives, other relatives and friends. This can help us understand, why Vassilikiot women frequently complain about the presence of their husbands in the coffee-house, but show remarkable patience with their husbands' intense involvement in hunting. Besides, the presence of men in the coffee-house is often delayed until late at night, a time when women cannot readily enjoy the company of other women and are destined to remain isolated in the home, in front of the television.[19] Furthermore, the coffee-house is associated with 'dangers' (*kindhynous*) related to potential involvement in gambling and excessive drinking – these activities being categorised by Vassilikiot women as a 'waste' (*spatali*) of the household's financial resources.

Contrary to the cordial, mostly co-operative relationship between local hunters, who are often neighbours or relatives, and who share, not merely an interest in hunting but also the same social and moral cosmos, non-local hunters in Vassilikos are treated with generalised antagonism. Some Vassilikiots take the opportunity to rent some of the hunting areas they control to outsiders, and thus make some profit. Exceptions to this are cases where the outsider is a relative or an individual to whom an obligation is owed. But even then, the greater community of local hunters, those who do not share these particular obligations, do not hesitate to express their competitive spirit towards non-local individuals who hunt in Vassilikos, especially those who come from urban centres outside the island.

The opposition between Vassilikiot hunters and outsiders becomes even more evident in the conflict over hunting prohibitions policed by the state authorities. As I have already noted, these prohibitions are never thoroughly enforced due to the determination of the local resistance and the indecisiveness of the state authorities; the attitude of the latter alternating 'between attempts at control and implicit tolerance' (Sutton 1998: 61). According to my observations, competition or discord with outsiders (hunters from the town or the authorities) further reinforces the solidarity of Vassilikiots at the local level. The patrols of the state authorities and the forestry department are met with collective excitement, as an opportunity for individual and collective performances of insubordination or, to quote Herzfeld (1985a: xii), 'defiant independence'. In this respect, hunting in Vassilikos resembles the 'explosive' custom of dynamite throwing in Kalymnos, with which the Kalymnians indicated their resistance to the Italian occupation (in the past) and the authority of the state (in more recent years) (Sutton 1996, 1998).[20] Similarly, the persistence with which Vassilikiots engage in illegal hunting further challenges the positions and ideals of conservationists and environmentalists who argue in favour of restricting the turtledove hunt. In the general context of the Vassilikiot opposition to environmental conservation and the establishment of the National Marine Park on the island, the articulation of a local pro-hunting and anti-environmentalist discourse has acquired a unique confrontational significance.

Unlike Icelandic whalers (Einarsson 1993) and North American Indians (Ellen 1986a) who have successfully championed their cause against conservationists by reference to arguments of the 'our-way-of-life' type, Vassilikiots do not systematically apply culturally-based arguments of that type to defend their hunting practices. They merely 'naturalise' (Yanagisako & Delaney 1995) hunting as an activity, which is an indispensable part of their life, 'something natural' (*kati fysiko*) for those 'who live in the countryside'. As has been well illustrated in earlier chapters, Vassilikiots perceive their farming way of life as a constant 'struggle' (*agonas*) with the natural environment and its animate or inanimate components. Wild animals or birds, which exist outside the context of 'care and order' established through the farmers' 'struggle', are not credited with the privileges and responsibilities that membership of a rural household entails. Hunting them is not constrained by the orderly cycles of life and death to which domestic animals are subjected, while their death directly provides the rural household with additional benefits: an extra meal which is simultaneously an ideal occasion for inviting guests, meeting obligations and celebrating sociality. The noisy, collective participation in hunting during the turtledove season is understood by its participants as a celebration itself. 'This is a celebration of life', one hunter – covered with dead birds hung all over his body for display – once told me and continued: 'We live here, we work here... and we have a great love for hunting, a great passion.'

NOTES

1 For further examples of poems as representative of expressive culture in the context of hunting, see Moore (1994: 85).
2 Blackbird, (*Kotsyfas*), *Turdus merula*. Hoopoe, (*Tsalapeteinos*), *Upupa epeps*. In Zakynthos it is

called (*Papouzas*). Cuckoo, (*Koukos*), *Cuculus canorus*.

3 Aristocratic women of the highly stratified Zakynthian society were famous for their elegant dress, which was always in touch with latest fashion in Europe. Their dress contrasted sharply – and produced equally sharp comments – with the way 'traditional' village women dressed.

4 Mistle Thrush, (*Tsikhla-Tsartsara*), *Turdus viscivorus*. Woodcock, (*Bekatsa*), *Scolopax rusticola*. Robin, (*Kokkinolaimis* and in Zakynthos called *Tsipourdhelos*), *Erithacus rubecula*.

5 See, also, Chapter Six.

6 '*An dhen eikhan poniria oi palioi tha khanosante. Dhen eikhan lefta gia fysegia. Pianan polla poulia me pagidhes*'.

7 Quail, (*Ortyki*), *Coturnix coturnix*.

8 Rumours say that some of the hunters have connections in the forestry department, or in the police headquarters, and are therefore in a position to know well in advance about an imminent inspection patrol.

9 'Τα τριγόνια λιγοστά, τα όπλα πολλά. Εγώ το ντουφέκι μου το πήρα δύο φορές, μετά το άφησα... Το μεσημέρι έφτασαν λίγα, ξελιγωμένα απ' την κάψα και τα χτύπησαν αμέσως. Αλλά χρόνια ήταν γεμάτοι οι ελαιώνες.'

10 '*Omos epeidhi sto nisi dhen ekhei alla kynigia, oi arkhes kanoun anokhi*'.

11 '*Tous therisan ta khimika*'

12 See the following magazines: 1. Κυνηγεσία και Κυνοφιλία. 2. Κυνηγετικά Νέα. Μηνιαίο κυνηγετικό περιοδικό του χθες και του σήμερα. 3. Κυνήγι και όχι μόνο. Περιοδικό κυνηγετικού – σκοπευτικού – φυσιολατρικού & κυνοφιλικού περιεχομένου. 4. Κυνήγι & Σκοποβολή. 5. Κυνηγός & Φύση. Το σύγχρονο περιοδικό για τους λάτρες της δράσης.

13 See previous footnote.'

14 It is worth mentioning, however, that the columnists and editors of the above mentioned magazines, although devoted hunters themselves, openly disapprove of the violation of the seasonal hunting prohibitions, and unlike the Vassilikiot turtledove hunters, they try to establish the new ethos of the lawful and 'responsible hunter' (*ypefthynos kynigos*).

15 '*M' ena smparo, dhyo trigonia*' [Killing two birds with one stone].

16 The older Vassilikiots remember an aristocratic woman from the island's capital, who frequently hunted on their land with a 'light' (*elafry*) gun, which was specially designed and manufactured for her. 'She was the only woman that ever hunted on our land', Vassilikiot hunters maintain and add, with a conspiratorial tone, 'you see, she was a lesbian and she didn't make the slightest effort to hide it; she was living in a big mansion with her girlfriend!' My older informants remark that 'this woman was the first woman ever to appear in Vassilikos wearing trousers' and unanimously attribute her preference for hunting to her 'male' tastes and temperament.

17 Seremetakis has recorded the only case in Greek ethnographic literature where women 'shared hunting with men' (1991: 44). In Inner Mani, Seremetakis explains, interclan feuding necessitated men spending lengthy periods of time shut up indoors, while women, who were treated as noncombatants, took responsibility for several outdoor labours, which are traditionally considered (in other parts of rural Greece) the primary duty of men (ibid.: 43–5).

18 Hunting as a symbolic expression of masculinity among Cypriot men is also recognised by Sheena Crawford (1982: 97). Her brief description of men's enthusiastic involvement with hunting in Kalavasos fits very closely with my own experience of hunting in Zakynthos.

19 The households in Vassilikos follow a widespread spatial pattern of settlement, and the community follows a long patrilocal tradition; the sense of isolation faced by Vassilikiot women differs markedly from the confidence and sense of solidarity experienced by women in the matrifocal neighbourhoods of Lesbos, studied by Papataxiarchis (1988, 1991, 1995).

20 It is also very interesting to note that Kalymnian women – in a way that much resembles Vassilikiot women's attitude towards hunting – do not appear to particularly oppose dynamite throwing 'despite the fact that dynamite throwing is the exclusive domain of men' (Sutton 1998: 62).

9
RELATING TO THE 'NATURAL' WORLD

৽৽৽৵

In this final chapter I bring together some conclusions drawn throughout the book regarding the cultural principles that inform Vassilikiots' resistance to ecological conservation. First, I shall discuss the confrontational, pragmatic spirit of Vassilikiots' engagement with the productive resources of their land, and the meaningfulness of this agonistic disposition for the local protagonists. Then I shall illustrate the relationship between this confrontational attitude and the indigenous worldview towards the natural world, reflecting upon the caring, but anthropocentric, attitudes that permeate Vassilikiots' relationship with animals and the environment, and the continual practices of labour or 'struggle' that inform those pragmatic attitudes.

To underline the relevance of the theoretical discussion that follows to the turtle conservation dispute considered in this book, I shall set out here the criticism most frequently articulated by the environmentalists against the protesting landowners in Vassilikos. The protectors of the sea turtle accuse Vassilikiots of being exclusively preoccupied with their personal interest and the maximisation of financial profit through the uncontrolled development of tourism. This accusation also contains an explanation. This is how the protectors of the sea turtle attempt to account for Vassilikiots' rigorous resistance to environmental conservation. The environmentalists, being unwilling or reluctant to challenge one of their fundamental axioms – namely, that the protection of natural species and the environment should be a top priority for all parties concerned – interpret the motivation of the protesters as ill will and aggression driven by personal economic interests. By placing disproportional emphasis on the materialist concerns of the Vassilikiots, the environmentalists, in their implicit social explanation, translate culture 'as an environment or means at the disposition of the manipulating individual' (Sahlins 1976:

102). The analysis that follows will demonstrate that such a one-dimensional inter-pretation is not merely reductionist, but also misleading (see, also, Theodossopoulos 1997a).

The logic of the Vassilikiots' struggle

The term 'logic' (*logiki*) in modern Greek conveys associations of reason or ratio-nality, but when used as an adjective for a particular practice, any practice, it becomes synonymous to meaningfulness or purposefulness. My use of the term 'logic' also alludes to Bourdieu (1977, 1990) who has underlined the ascription of social meaning to repetitive practices of labouring in fields of everyday action. For my friends and respondents in Vassilikos, tourism and farming are such fields of action. Their 'practical mastery' in these fields – their expertise in 'the specific regularities that constitute the economy' of tourism and farming – becomes 'the basis of sensible practices', that is practices with logic (*logiki*) (Bourdieu 1990: 66). For all the Vassilikiots I know, it is indeed imperative that their daily work is meaningful and that 'their struggle' (*o agonas tous*) in various fields of toil has logic or purpose.

The notion of 'struggle' (*agonas, pali*), as indicative of a more general combat-ive attitude towards life, and in particular as an experience emanating from the constant embodiment of physical fatigue during work, has been registered ethno-graphically in several anthropological accounts of Greek communities (Friedl 1962: 75; du Boulay 1974: 56, 1986: 154; Kenna 1990: 149–50; Hart 1992: 65–6; Dubisch 1995: 215; Argyrou 1997: 163). This agonistic disposition towards life often characterises 'the agonistic quality of human relations' in the local society (Friedl 1970: 216), but becomes explicit in the indigenous relation-ship with the natural world. The connotations of effort or contest raised by the notion of 'struggle' demarcate the immediate environment, or even the limits of one's stamina, in terms defined by an endless, continual confrontation. As Argy-rou (1997: 163) points out, 'out of this confrontation – akin to physical combat – the world emerges as a formidable adversary and the Self emerges as a physically and mentally strong individual who, far from being deterred by the challenge, welcomes and even provokes it.' In this respect the confrontational spirit of the 'struggle' itself provides individual actors with the impetus required to face the repetitive strain of those practices that are locally conceptualised as work.

In Vassilikos the repetitive yearly cycles of agricultural labour are comple-mented by work devoted to the growing economy of tourism. Most men and women are involved to one degree or another with both tourism and farming and maintain that work invested in both those fields constitutes 'struggle'. Their 'struggling' practices are mostly enacted with in the physical setting of family 'property' (*periousia*), that is, on Vassilikiots' land and the structures erected on it (farm buildings, tourist establishments), which comprise the fertile common ground for the realisation of both tourism and farming. For example, in the olive groves, which require some care and attention all year round,[1] tourist apartments, which also need maintenance during and after the periodic visits of tourists, are built. Similarly, the local products of cultivation or animal husbandry are often put to use by the producers themselves in their own tourist enterprises. In this context of relative affinity between different types of labour investment, tourism, cultivation and animal husbandry are locally conceptualised as available resources of the land that should not be wasted. This is the logic of self-sufficiency proper (du Boulay 1974: 244, 247; Loizos 1975: 41, 50; Kenna 1976b: 349–50, 1990: 151–2, 1995: 135, 2001: 32; Just 2000: 203; see also, Gudeman & Rivera 1990: 44–5) – also referred to by other authors as 'autarky' (Stewart 1991: 60, 65) or 'subsistent prototype' (Pina-Cabral 1986: 32) – an analytical construct employed by anthropologists to acknowledge an indigenous principle that renders meaningful a great deal of indigenous labour.

In Vassilikiot everyday subsistence strategies self-sufficiency resonates with a practical code, a rule of thumb that enforces the maximisation of all the subsistence resources one's land can provide. These involve the recycling of old materials on the farm, the use of locally produced farming goods in tourism, but also the recruitment of all available family labour as opposed to employing outsiders. In this sense, self-sufficiency informs meaningful practices enacted by members of independent households engaged in the perpetual 'struggle' of safeguarding and advancing the welfare and prosperity of their own household. Here, self-sufficiency is intimately related to family interest (*symferon*), an indigenous concept, to which Vassilikiots refer by the same term used in modern Greek for denoting self-interest. This tautology helps us understand that in the indigenous discourse self-interest primarily means the welfare of households, as opposed to the personal benefit of particular individuals. As Hirschon points out, 'the notion of the single individual as it is understood in the West may still be inappropriate for understanding contemporary Greek society' (1989: 141) and it was indeed inappropriate for understanding the economic practices of the great majority of Vassilikiots in the 1990s.

The family-oriented, corporate understanding of self-interest (cf. du Boulay 1974: 169–70; Loizos 1975: 66, 291, Hirschon 1989: 104, 141, 260), sets the parameters of purposefulness in Vassilikiots' daily work or 'struggle'. For example, in Chapter Five, I examined how Vassilikiot women willingly participate in agricultural endeavours, such as the olive harvest, which involve uncomfortable manual labour in the open. Drawing upon Strathern's (1988) *The Gender of the Gift*, I explained Vassilikiot women's participation in the harvest not merely as a contribution to their household's self-sufficiency and prestige, but also as a statement of their intention to invest in meaningful relationships with their husbands and other members of their households. Similarly, in the field of tourism, Vassilikiot men and women of all generations unite their forces to deal with work that involves different skills and talents. Young Vassilikiots with a better command of foreign languages deal with the tourist customers in the local family restaurants, while older relatives cook at the back stage. Vassilikiot women 'clean and maintain' the tourist apartments, while Vassilikiot men 'clean and maintain' the agricultural landscape within which the apartments are built. In all cases, self-interest is defined in terms of household prestige or financial success, and in all cases, Vassilikiots depend upon, and explicitly articulate their acknowledgement of, the co-operative 'struggle' (*agona*) of their spouses, parents or children. 'The successful running of the household brings prestige' to all parties involved (Loizos & Papataxiarchis 1991a: 8), but also further underlines the meaningfulness of the labour investments made by all.

Bearing in mind that Vassilikiots understand the purpose of their involvement in tourism in terms of their generalised daily struggle to enhance the welfare and prestige of their household, it becomes much easier to appreciate their frustration and anger at the regulations aimed at environmental conservation. To the extent that conservation inhibits Vassilikiots' investment in the tourist economy, it renders their invested labour and effort meaningless. The Vassilikiots themselves refer to this unfortunate circumstance as an experience of loss; 'our toil is lost', they repeatedly maintain, 'the ecologists have done us great harm, now our effort is wasted.' The implementation of the conservation law in Vassilikos has prevented some local families from developing tourism on their land. These restrictions render the most vital, available, productive resource of this particular land unavailable, bluntly violating the indigenous notions of self-sufficiency and self-interest. In other words, conservation does not merely harm the financial prospects of some indigenous actors; it contests the logic or meaningfulness of their 'struggle' in life, a logic embodied, in Bourdieu's (1990) terms, in the practice of struggle itself.

Finally, the accusation raised by the environmentalists in Zakynthos concerning the amoral individualism of the local tourist entrepreneurs loses its force when put against the context of Vassilikiots' daily economic practices. The latter are very well prepared to support the moral impetus of their struggle. As Parry and Bloch (1989) point out, in several societies prolonged involvement in the short-term, monetary sphere of exchange could be considered a threat to the long term wellbeing and moral order of the community. In Vassilikiot social life, the short-term and overtly monetary character of transactions in tourism, is mediated, and reconciled with long-term social expectations. This process of transformation (Parry & Bloch 1989: 23–7; cf. Toren 1989; Carsten 1989) is realised by Vassilikiots' constant evocation of the family-oriented motivation of their economic ventures. What the tourist entrepreneurs in Vassilikos clearly and repeatedly stress is that their investment in the tourism industry is their contribution to the future of their family, the marriage of their daughters or sons. In fact their daughters and sons willingly participate in family tourist enterprises, a labour contribution which fortifies the self-sufficiency and self-interest of their households, but also, further enhances the claims made by the younger generation to these productive resources. When Vassilikiots point out that, unlike the urban conservationists who rely on other sources of income to save the environment and advance their careers, 'the people on this land have daughters to marry' (cf. Herzfeld 1991a), they allude to cultural priorities that go far beyond the level of monetary transactions.

A working relationship with the land

The setting of Vassilikiots' confrontation with the natural world is the land of Vassilikos. In the past, powerful landlords controlled the most fertile lands on the island and the majority of the rural population had to put up with strict patterns of economic exploitation (*sempremata*) regulating the tenancy agreements (*sempries*) between labourers (*kopiastes*) and the wealthy 'masters' (*afentadhes*) of the land. In the years following the Second World War the landless tenants (*semproi*) of Vassilikos, gradually, but steadily, acquired small plots of land of their own, which they nowadays proudly consider to be the result of years of hard labour and toil. For the older generations of Vassilikiots, those men and women who witnessed and participated in the gradual and difficult process of land acquisition, land ownership signifies the materialisation of their own and their parents' efforts, 'their struggle' (*ton agona tous*) which is inscribed on the land itself. The younger Vassilikiots, who were lucky enough to escape 'the fate of the landless tenant' (*tin moira tou semprou*), despite occasional criticisms of their parents' back-breaking involvement with relatively unprofitable cultivation, similarly appreciate the value of Vassilikiot land, which has now acquired an additional significance as the solid foundation for profitable tourism development.

It must be noted, however, that the cultural significance of the land in Vassilikos extends beyond the narrow conceptual calculation of its material value. Even those interactions with the land 'generally presented as simply 'economic' are as heavily symbolic and expressive as any other cultural form' (Strang 1997: 83). 'Land has more than purely economic uses', Davis (1973: 73) has argued, and these are, in fact, well attested in the anthropological literature on Mediterranean country-people. Land influences marriage strategies, strengthens ties of unity of households, and constitutes an imperishable part of a household's history and collective identity (Lison-Tolosana 1966, du Boulay 1974, Pina-Cabral 1986). It also signifies self-sufficiency, security, status, political influence, and the independence of household members – especially female ones – from the stigma of paid labour (Davis 1973, Loizos 1975). A working relationship with the land is synonymous with responsibility, power, vitality and good health, and it is a major prerequisite for establishing a local identity (Pina-Cabral 1986: 25, 67, 152–3, 126, 208). In Vassilikos a bond with the land – such as land ownership or a working engagement with it through a period of several years – raises claims to participation in the tourism economy and legitimises the pursuit of economic benefits in tourism.

Hence, the landowners affected by conservation in Vassilikos feel that they legitimately deserve to develop tourism on their property. Their ownership of land and the labour they have invested in it justifies this claim. A Vassilikiot in his forties, who is involved with more than one tourist enterprise, puts this into perspective: 'It is unfair for a hard-working man (*dhouleftaras*) like my cousin to be prevented from prospering because of the turtle(s)…he just deserves to build as much as everyone of us.' Taking this view one step further, Vassilikiots argue that any hard-working individual 'deserves' (*axizei*) to be rewarded for his or her toil or struggle. Comparisons with neighbouring regions on the same island, where tourism has made a drastic improvement in local people's economic circumstances, only amplify Vassilikiots' suspicions that the conservation of their land is a well orchestrated conspiracy, 'a great injustice' directed against them.

In their efforts to justify their claim to control over the land under conservation, the protesting landowners unite several themes of land valorisation in a unified discourse that aims to highlight the value of their land. But while environmentalists interpret this emphasis on land valorisation as a materialistic calculation of profit, the Vassilikiots are attempting to put across the opposite message. They accentuate the cultural signification of their land in their attempts to demonstrate the depth of their symbolic relationship with it and their unwillingness to be parted from it. In their arguments, the frustration of waiting in vain for compensation blends with memories of landlessness in the past, while their struggle to develop tourism appears inseparable from their parents' battle to acquire some land of their own. As I mentioned in Chapter Three, polysemy in this context brings about certainty rather than confusion; when divergent sets of value point in the same direction – namely, that land is a very serious asset for the landed actors – the local protest against conservation measures obtains the moral quality of a just cause (Theodossopoulos 2000).

As has already become apparent from my reference to other studies throughout this book, anthropology has contributed a great deal to the recognition and acknowledgement of the multiplicity of cultural meanings ascribed to land – at least within the Greek and the Mediterranean ethnographic contexts. The sensitivity of earlier generations of scholars towards the ethnographic particulars of the relationship between landed actors and their land has recently been complemented by a new academic interest in landscape (Bender 1993a, Tilley 1994, Hirsch & O'Hanlon 1995, Schama 1995). Landscape studies in anthropology have facilitated a better conceptualisation of particular gazes or ways of looking at the physical world, viewpoints saturated with meaning. Thus, it is not difficult to realise that all the multiple sets of land valorisation described so far emanate from different aspects of the human-environmental relationship and reflect different readings and experiences of the land. The turtle breeding area in Vassilikos, for example, does not merely comprise an 'empty space, untouched and wild' (Strang

1997: 210), neither one filled only with turtles. It contains polysemic significations of the land, personal histories of labour and struggle embedded in it (Bender 1993b). Landscape theory helps us appreciate that when different readings of the land clash in the context of an environmental dispute, what is disputed is not one 'space, only spaces' (Tilley 1994: 10); not one landscape, but 'many landscapes in tension' (Bender 1993b: 2); not a single, one-dimensional way of referring to the value of Vassilikiot land, but several.

Land, however, unlike landscape, relates to material and tangible dimensions of economic life that stretch beyond the conceptual and experiential realm of the visual (Abramson 2000). This is why in the ethnographic sections of this book I devoted some considerable attention to land tenure and the economy of the land, 'not losing sight of the different regimes of meaning in which patterns of property in land crystallise' (ibid.: 2). More importantly, I have tried to highlight the practical engagement of the landed actors in Vassilikos with their immediate physical environment, their pragmatic immersion in the reality of work. In Vassilikiots' human-land relationship there is a primacy of the practical over the aesthetic, and this indigenous predilection justifies my prioritisation of the term 'land' over the term 'landscape' in my analysis. Hirsch has defined the landscape as a process, that is the articulation of a 'foreground of everyday social life' with a 'background potential social existence' (1995: 4–5, 22–3). Vassilikiots' attitude towards their immediate physical environment is primarily focused on the foreground actuality of their work, their struggle to render the resources of the land productive. It is only in terms of this emphasis on the actuality of work that the background potentiality of Vassilikiots' land becomes meaningful. At the end of a hard day's work the tired labourer will stare at the freshly ploughed soil of a cultivated field or the recently renovated tourist apartments situated in it and murmur: 'This is how I want this land to be, clean and well-cared for.'

Vassilikiots' perception of the environment is shaped 'by the practices and activities going on within it' (Green & King 2001: 285), and by the Vassilikiots' own toil and struggle. 'Houses, fields, graves' and so many other physical indications of 'rights and obligations' (Kenna 1976: 21) feature as prominent markers of the human investment in the local landscape. The working endeavours of Vassilikiots unfold in space and time, permeating the calendar of working activities with agricultural tasks (Hart 1992, Greger 1988) and labour invested in tourism that expands beyond seasonal considerations (Abram & Waldren 1996). Their constant involvement in both kinds of activities (agriculture and tourism) ensures their continual caring presence on their land. The natural world in Vassilikos is not conceptualised as the numerical aggregate of its animate and inanimate components. Its meaningful existence is defined in terms of the work invested in it by its human caretakers, the inhabitants of Vassilikos who 'live in' or 'dwell' in it (Ingold 1995: 58) and interweave their lives with it (Green 1997: 638). By the term 'nature' (*fysi*), Vassilikiots do not merely designate the uninhabited wilderness that surrounds their farmland. The well tended cultivated fields that stretch along the main village road are also referred to as 'nature'; a part of the local environment that is considered to be beautiful because it embodies the toil of the people who care for it.

'Look around you', my friends in Vassilikos maintain, 'those olive groves, the gardens, our fields are beautiful nature'. And they add after a short reflective pause: 'Nature needs care and this is why we struggle every day; we work the land and care for it.'

Anthropocentrism, order and care

In the context of Vassilikiots' everyday struggle with the productive resources of their environment the indigenous concept of 'order' (*taxi*) addresses the establishment of control (or sense of control) over the living and constantly regenerating parts of the natural world. An example of this is the Vassilikiots' endless battle against undesirable vegetation that grows continuously on their farmland and the spaces proximate to their tourism enterprises. But nowhere else is this vigilant safeguarding of order more clearly manifested than in the relationship of Vassilikiots with their domestic animals. Despite the recent, successful introduction of tourism to Vassilikos, the great majority of the community's inhabitants keep animals on their farmland and care for them devotedly. Although men usually tend the larger animals and women the smaller ones, those sheltered in proximity to the domestic compound (cf. Friedl 1967: 103–4; Handman 1987: 151–2; Hart 1992: 243–6; Galani-Moutafi 1993: 254; see also, Pina-Cabral 1986: 83), most members of Vassilikiot households contribute with their labour to the daily care of 'their animals' as best they can.

Domestic animals in turn are expected to reciprocate the care they have received by contributing to the welfare of the household to which they belong. As du Boulay (1974: 16, 86–89) has pointed out, domestic animals in rural Greece are perceived as members of the household, albeit at the lowest level, and their membership entails their receiving and providing services. Reciprocity in this case involves responsibilities, such as the obligation of respecting the order of the farm, and rights, such as the provision of food, shelter, medical care and protection from predators. According to this perspective, the organised slaughtering of domestic animals is interpreted as a contribution to the prosperity of the farming household – a necessary sacrifice for the maintenance of the remaining animals on the farm. By contrast, the sudden and unpredictable death of domestic animals from predation or illness is treated by Vassilikiot farmers as a source of sorrow, a waste of all the daily struggle invested in the care of these animals.

In their relationship with their animals, Vassilikiot farmers assume the role of the caretaker (cf. Palsson 1996: 71) and provider, and maintain that without their caring presence farm animals will lapse into disorder and perish. This is why they persistently punish those animals that transgress the order of the farm as defined in human terms and reward those who learn their place or position in it. When punishing their animals they often talk to them, scolding them like children or comparing their misbehaviour with the orderly and more obedient conduct of some other animals. They expect them to learn simple routines related to their feeding and sheltering, thus saving their owners from some additional effort during their daily 'struggle' to take care of them. As far as I observed, most farm animals learn what is expected from them and read the most obvious of the farmer's intentions in the farmer's body language and intonation. Vassilikiots derive satisfaction from this reciprocal interaction, arguing in favour of their involvement with animal husbandry despite its limited material rewards.

Unlike animals living on the farm, the wild creatures of the Zakynthian fauna are treated with either indifference or hostility in Vassilikos. Those in a position to harm domestic animals or cultivation are hunted down persistently (cf. Moore 1994, Marvin 2000), while the ones that do not directly threaten or encounter the human protagonists do not attract the attention of the Vassilikiots and rarely figure in conversation. The sea turtle is the only exception, a once inconspicuous wild creature which has lately caused – because of the restrictions imposed for its conservation – a great deal of concern in Vassilikiot life. Sea turtles are remembered by the older men and women in Vassilikos as creatures of no particular importance, animals without any apparent use. Nowadays Vassilikiots almost unanimously maintain that 'the turtle' has caused 'great harm' to the people. In this respect, the obstacles to tourism development caused by turtle conservation – and the negative consequences this has for the prosperity of some Vassilikiot households – are treated as damage very similar in kind to the 'harm' caused by the wild predators that violate the order and safety of the farm.

There exists a third category of wild creature, one that receives 'more attention from' (Marvin 2000: 206), and excites the imagination of, most Vassilikiot men. It consists of those animal species that men hunt, in the case of Vassilikos, primarily birds. Hunting in Zakynthos is firmly rooted in local culture and recognised as a celebrated passion (*pathos*). Vassilikiots indulge in it and defy the seasonal hunting regulations in an uncompromising manner – treating it as a further opportunity to declare their opposition to environmental conservation. During the spring turtledove hunt, the local hunters display their disregard for the hunting ban in stances of 'performative excellence' (Herzfeld 1985a: 16), shooting at the passing birds at any given opportunity. As I described in Chapter Eight, the hunters themselves are unhappy with the decline in the game, but appear unwilling to interrupt or even inhibit their hunting passion. Wild animals are not 'engaged by the structure of social relations of the human community' and are therefore viewed by Vassilikiot hunters as 'living resources' of the local environment (Ingold 1986: 113). Unlike domestic animals, they exist outside the context of care and order established on the farm and are not bound to the human protagonists by any reciprocal ties.

It is apparent that household-centred priorities, like those that inform most aspects of Vassilikiots' working engagement with their environment, outline the normative and practical dimensions of the human animal relationship in Vassilikos. The local perception of order – and the care (or lack of care) towards natural creatures that emanates from it – is intrinsically dependent upon the well-established central position of the human protagonists in 'their' local environment. Considering anthropocentrism as the tendency to approach, understand, classify and treat animals as beings peripheral to a centrally positioned human self, or as beings existing in order to serve and satisfy human needs, it is fair to label Vassilikiot attitudes to wild and domestic animals as anthropocentric (cf. Papagaroufali 1996: 244; Theodossopoulos 1997a: 263).

Vassilikiots' anthropocentric – or human household-centred – attitude towards the animate and inanimate physical environment is supported and ideologically reinforced by an elaborate religious cosmology. The anthropocentric orientation of Christianity in its approach to the natural world has been emphasised by several scholars (White 1968, Worster 1977, Morris 1981, Thomas 1983, Serpell 1986, Ritvo 1987, Ingold 1988, 1994, Tapper 1988, Willis 1990, Davies 1994). My detailed analysis of St Basil's taxonomic insights, in Chapter Seven, offers a clear view of the dogmatic cosmological priorities with respect to the animal world set up by the Orthodox tradition. In St Basil's classificatory discourse the primary taxonomic principles regarding the higher and more inclusive levels of animal and plant categorisation are provided by the Bible. Sharp categorical distinctions separate plants (which are considered inanimate) from animate beings; swimming and flying animals (which have life, but a somewhat imperfect form of life) from animals of the land (which are superior living creatures); and all non-human living creatures from human beings (which are made in the image of the Creator). However, this hierarchical form of animal classification is dramatically emancipated from its dogmatic guidelines at the lowest, less inclusive, taxonomic level. At this 'basic level' of categorisation St Basil relies upon empirical classificatory criteria, such as morphological characteristics, behavioural patterns, means of procreation, habitat, nutrition and lifestyle (Bulmer 1967: 6, 1970: 1072–3; Atran 1990: 214, 5–6, 29, 56 1993: 57–9, 64; Ellen 1993: 93–4).

This discrepancy between normative and practical/empirical considerations at the higher and lower levels of classification is similarly reflected in the human environmental interaction in Vassilikos. Vassilikiots' attitudes towards the animate and inanimate environment is informed by a coherent cultural worldview with roots in religious cosmology that helps them define the parameters of care and order and the priorities in the management of their farmland. But this cultural background does not always provide Vassilikiots with ready-made solutions in their everyday interaction with their environment. It only suggests some conceptual principles regarding the position of the human protagonists in a physical setting that includes wild and domesticated animals, cultivated fields and wilderness. In their daily practice Vassilikiot farmers constantly have to improvise while experimenting with cultivation and animal husbandry and apply their own empirical solutions to problems related to their human-environmental interaction (Richards 1986, 1993, 1996; Croll & Parkin 1992b). Thus, their conceptualisation of order in the farm environment and the hierarchical distinctions between human and non-human beings that this order implies are closely dependent upon the realisation of caring for the animate and inanimate environment and the continual work or struggle this care entails.

In the anthropological literature on human attitudes to animals, we can easily identify a distinct contrast in attitudes towards animals between agricultural societies and hunter-gatherers. The agriculturalists are prone to exhibit antagonism towards the natural world, attempting to dominate and control it, while hunter-gatherers usually treat animals and nature in a more egalitarian way (Ingold 1980, 1986, 1994). Morris recognises this contrast between hunter-gatherer and agricultural 'cultural attitudes' to animals and further acknowledges that the farming way of life has an 'antagonistic' orientation towards animal life (1995: 303–4; 1998: 3–4). However, he is sceptical about the abrupt division of diverse cultural attitudes towards animals into two rigid categories: the pre-literate cluster of societies with the 'egalitarian, sacramental' viewpoint of nature, and the Western cultural traditions characterised by a mechanistic, dualistic and controlling approach towards the natural environment (Morris 1995: 302–3; 1998: 1–6). 'Many scholars', he argues, 'write as if historically there are only two possible "world-views", the mechanistic (anthropocentric) and the organismic (ecocentric)' (1995: 303). This generalising tendency obviously underestimates the diversity and changing character of Western traditions – which includes a multiplicity of different ontologies and historically specific understandings of nature (cf. Palsson 1990, Strang 1997)[2/3] – and fails to account for particular cultural practices, such as that of the Vassilikiots, where the two contrasting attitudes often coexist and complement each other (Morris 1995: 301–12; 1998: 2).

Beyond the contrast between agriculturalists' and hunter-gatherers' attitudes to animals, a second generalising distinction can be identified between the 'primitive, archaic, tribal or premodern' cluster of cultures on the one hand, and the 'modern, Western' world view on the other (Willis 1990: 20; Ellen 1996b: 103). The latter has been generally associated with utilitarianism and anthropocentric hierarchies which are presumed to be opposed to the ecologists' and 'tribal' people's *balanced, reciprocal, interdependent, holistic* approach to their natural environment. My own ethnographic account clearly depicts the ethnocentric disposition of this distinction. The traditional relationship of the people in Vassilikos with animals is permeated by a combination of pragmatism and care, a mixture of 'Western' anthropocentrism and 'pre-modern' reciprocity. Evidently, their attitudes towards, and their everyday interaction with, animals hardly fit into generalising categories of that type.

Willis (1975), for example, has compared attitudes towards animals from three African examples: Evans-Pritchard's Nuer, Mary Douglas's Lele and the Fipa, agriculturalists in south-west Tanzania, studied by himself. He remarks that the Nuer, well-known for their attachment to their domestic cattle, dislike wild animals, while the Lele regard their domestic animals with disdain or contempt and are much more positive about hunting, an activity they invest with prestige and mystical meaning (ibid.: 44–6). But Fipa attitudes to both wild and domestic animals is described by Willis with the terms: 'utilitarian', 'irrelevant', 'neutral', 'businesslike' and 'down to earth' (ibid.: 45–50). 'What is the use of that to us, the human community?', the Fipa wonder when confronted with animals and objects of the external world (ibid.: 50), and their unashamedly pragmatic evaluations closely resemble my own respondents' bewilderment as to the 'use' of turtles and monk seals, species that are protected by conservation legislation in Zakynthos. In this respect, both Fipa and Vassilikiot pragmatism contrast sharply with the idealised ecological depictions of pre-modern world-views.

Even in the Western European tradition, where attitudes towards most animals have been primarily economic and exploitative, there is a notable exception: pets and pet-keeping, a subject studied thoroughly by James Serpell (1986, 1989, Serpell and Paul 1994). Pets are by definition animals loved for 'no obvious practical and economic purpose' and, as Serpell persuasively argues, sheer material utility is not a valid model for explaining the human tendency to keep pets (1986, 1989). Serpell's work provides further evidence that cross-cultural categorization of human attitudes to animals according to *utilitarian – Western* versus *non-utilitarian – traditional* dichotomies is untenable. Serpell demonstrates methodically and by use of abundant examples that pet-keeping is widespread in numerous pre-modern societies. In some of these societies animals are treated in a strictly utilitarian manner but, at the same time, some of them – even animals of species which are in general mistreated – are kept as pets, independent of any material considerations (1986: 56–7, 1989: 13). In this sense, pre-modern or traditional societies are not markedly different from the 'Western-modern' ones: utilitarian attitudes to animals and unconditional care sometimes co-exist within the same culture, the same village, or even the same farm.

Despite his acknowledgement of the complex and contradictory character of human-animal relationships, Serpell attempts some historical reflections of a generalising nature. He demonstrates with several examples that ancient Greeks and Romans, at least in most cases, approached nature as 'a fearsome opponent to be mastered and avoided' (1986: 175–7). The Aristotelian natural hierarchies, and Plato's emphasis on the power of human reason, were historically succeeded by Christian anthropocentrism and the biblical notion of human 'dominion over every living thing' (Serpell 1986: 122–3, Serpell and Paul 1994:132). But Serpell also notes that within the Christian world view, with its emphasis on human superiority and animal subordination, several exceptions could be identified, such as the friendly attitude to animals exhibited by St John Chrysostom, the Franciscan Order and, even the medieval heresy of the Cathars (1986: 122,126). Similarly, Morris (1981: 131–2) and Ellen (1996a: 13) underline the anthropomorphic, animistic perceptions pre-Christian Greeks had of the natural world, as opposed to other scholars who emphasise the hierarchical and anthropocentric. This is a further example of the dangers underlying both historical and cross-cultural generalisations. Morris, is stressing the holistic, animistic world-views of Plato, while Serpell (1986) and Thomas (1983) are crediting the ancient philosopher with enhancing the dichotomy between man and animals with his veneration of human reason.

In a similar manner Keith Thomas (1983) observes that anthropocentric perceptions of the natural world are not merely confined to the Judaeo-Christian tradition. He distinguishes, for example, between the rather 'ambivalent' attitudes of the Christian religion, oscillating between 'domination' and 'responsibility' towards non-human beings, and the evidently anthropocentric – and often religious – orientation of several individual scholars in the early modern period (sixteenth to eighteenth centuries) (1983: 23–4). Harriet Ritvo (1987, 1994), focusing on a period succeeding the one studied by Thomas, unravels the complicated, almost incoherent character of Victorian attitudes to animals. Vivid examples of this chaotic multiplicity of views and information are portrayed in Ritvo's account of colonial hunters narrating stories to their Victorian armchair audiences about subjugating wild exotic beasts, or the eighteenth to nineteenth century bestiaries 'echoing anthropocentric and sentimental projections' of animal characteristics and dispositions: the 'noble' horse, the 'vicious' boar, the 'docile' elephant! (1987: 7–30, 1994: 113–115) In western European society, categorising and describing animals according to distinctions such as, '*edible-inedible, wild-tame, useful-useless*' – an approach often followed by my own respondents in Vassilikos – was gradually succeeded by a growing concern for 'systematic' classification, a commitment undertaken by specialists, the 'naturalists'. But despite the criticism and contempt of the 'naturalists' for unsystematic bestiaries and folk-taxonomies, natural history, like the earlier, religious versions of anthropocentrism, placed humanity at the apex of the newly founded classificatory hierarchies (Ritvo 1987: 13–4, 1994: 115).

Concluding this short examination of human attitudes to animals and the natural world, I suggest that anthropocentrism and anthropomorphism, antagonism and veneration of animals and nature, can hardly be confined to general categories spanning broad historical periods and cultural-regions, and can sometimes hardly be distinguished in the world-view of particular cultural traditions or within the writing of particular authors. Attempts to categorise different attitudes towards animals according to large clusters of cultures using terms like *modern, traditional* or *western-European* are in general unsuccessful, and the terms themselves are equally misleading. None of them can be accurately applied to the community I studied in Zakynthos and to its modern European, but still ambivalent about their western identity (Herzfeld 1987), inhabitants. The most serious objections to those terms will come from my respondents themselves, since most of them frequently shift their rhetorical 'self-definitions' from one category to another with surprising ease and exhilaration. Vassilikiots are Europeans living in a modern era, aspiring to some modern comforts, and faithfully adhering to several received values. Their relationship with animals and the natural environment reflects practical considerations, arising from their engagement with farming and tourism, in the fields of everyday work or 'struggle'.

A final note on conservation and controversy

As I have highlighted in the beginning of this chapter, the environmentalists who promote the protection of the sea turtles and the natural environment in Zakynthos describe those Vassilikiot landowners who protest against environmental conservation as calculating individuals, motivated primarily by personal interests. This negative portrayal, which is explicitly put forward as an accusation, also contains an interpretation of the indigenous resistance to conservation. The Vassilikiots are presented as too much preoccupied with the maximisation of their personal profit to take proper account of considerations related to the protection of their environment. Examined from an anthropological perspective, however, this negative portrayal of the Vassilikiots would be justly considered as an example of utilitarian functionalism (Sahlins 1976) or 'vulgar materialism' (Friedman 1974), a one-dimensional interpretation which systematically underplays the cultural and symbolic signification of the human-environmental relationship in Vassilikos. Throughout the chapters of this book, I have indirectly addressed this reductionist explanation by presenting numerous examples demonstrating the richness of Vassilikiots' relationship with the natural world.

In the preceding sections of this chapter, I have discussed the economic logic embedded in Vassilikiots' work or struggle and some of the most fundamental aspects pertaining to their worldview as regards the natural environment. I have underlined Vassilikiots' emphasis on the family-oriented priorities of their economic strategies and the meaningfulness of their daily work or 'struggle', which is most often realised through the close co-operation of family members in the fields of tourism and agriculture. In this respect, I challenged the presuppositions of the environmentalists regarding the selfish and individualistic motivations of those Vassilikiots' who protest against conservation. Men and women in Vassilikos perceive their engagement with the productive resources of their land – tourism, agriculture, animal husbandry – as an investment in the wellbeing and the future reproduction of their families. Their 'long-term' expectations inform their 'short-term' transactions in their daily economic activities (Parry & Bloch 1989), while their aspirations for the future are embodied in – and are in fact, inseparable from – the enactment of repetitive labour in the present (Bourdieu 1990). As Herzfeld (1991a) has illustrated in his account of a conservation dispute in Crete,[4] the property owners who are affected by conservation have daughters (and sons) to marry. This is an indigenous cultural priority that the conservationists (who in the Greek context are usually outsiders) appear reluctant to take into serious consideration.

Similarly, those individuals and pressure groups that advocate the cause of the sea turtles in Zakynthos, portray the Vassilikiots as people disengaged from caring for the natural world and the non-human creatures living in it. The Vassilikiots, on the other hand, maintain that all the daily work they invest in their land is a form of care (*frontidha*) which is enacted with constant struggle and devotion. 'We are working with the land every day', they rhetorically declare, 'we take care of the animals, the fields and everything that you see in Vassilikos and it is beautiful.' Keeping the animate and the inanimate environment in order is for Vassilikiot men and women the ultimate manifestation of care. To speak in terms of the Vassilikiots' interaction with the environment, between 'nature' (*fysi*) and 'culture' (*politismos*) or 'society' (*koinonia*), lies a large intermediate terrain, the 'cultivated land', which is at the same time part of both 'culture' and 'nature'. By replacing the abstract word 'culture' with the local term 'community' (*koinotita*) or 'village' (*khorio*), the cultivated land can be understood as 'cultivated nature' (*kalliergimeni fysi*), an extension of the village itself. Thus, 'nature' in the context of Vassilikos, can be understood as a more inclusive, non-basic folk classificatory category, and as such, it 'cannot be objectively defined, and no firm distinction between perceptual and social can be sustained' (Ellen 1996b: 118). More importantly, the Vassilikiots, far from being uninterested in the natural world, are continuously involved in caring for it. As the ethnography presented in this book has demonstrated, their caring practices are informed by a rich, culturally embedded human-environmental perspective that 'embraces not only the world of *anthropos*, but also that part of the world with which humans interact' (Descola & Palsson 1996: 14).

In the context of the environmental dispute, however, the 'thick' – to employ Geertz's (1973) overused expression – cultural signification of the indigenous relationship with the environment is systematically underplayed. Despite the plethora of cultural meaning attributed to the Vassilikiot land, both the conservationists and the State expect the Vassilikiots to surrender their claims to it without any compensation. This situation is perceived by the local landowners as a paradox or injustice. Determined to oppose environmental conservation, they defy its application at the local level, unable to challenge the conservation law itself. Their painful confrontations with conservationists, bureaucrats and state officials, resonate with a broader, recurrent theme in the regional literature: indigenous insecurity, inadequacy, mistrust and frustration in formal and informal dealings with representatives of the State and the law (Campbell 1964; Loizos 1974; Herzfeld 1991a, 1992). When it comes to direct representation in the process of presenting indigenous rights to the wider community, local actors remain – to use Ardener's (1975: 21–2) term – 'muted' (Theodossopoulos 2000: 73). Despite their articulacy in local conversation, Vassilikiot men and women speak neither the formal language understood by the agents of the state, nor the 'environmentally sensitive' prose popular in the media. In their struggle against environmental conservation, they only resort to strategic, but ephemeral, resistance to the regulations made by others for them.

Misunderstanding, distrust and hostility between those who impose conservation measures in Zakynthos and those whose lives are directly affected by them, has resulted in the long-running environmental dispute which has served as the common thread uniting the themes examined in this book. The human-environmental relationship espoused by the protectors of the sea turtles in Zakynthos and the human-environmental relationship embraced by the Vassilikiots who protest against sea turtle conservation, are each respectively informed by a cultural perspective, 'a particular way of understanding the world' (Milton 1996: 33). In fact, both parties involved in this environmental dispute, despite the differences in their respective cultural perspectives, maintain that they care about the environment.[5] The caring attitude of the environmentalists is conspicuously advertised. But the caring potential of the Vassilikiot human-environmental relationship, hidden within the 'thickness' of the indigenous culture and concealed by the muted positionality of its indigenous authors, has been communicated outside the confines of the local community for the first time through the pages of this book. This is why I wholeheartedly apologise to the people of Vassilikos for taking so much time to complete it. They will have the last word:

> When we were children there were masters (*afentadhes*), big and small landlords. They used to tell you, 'do that', 'don't do this'. Nowadays, we have the ecologists. They come and tell you, 'don't hunt', 'don't build', 'don't kill your own animals'! It is because of the turtle, they say...

> Look at those fields around you. Who cares about this land? Man (*o anthropos*) has to care for the world around him, to maintain it. Caring about the land and the animals is hard work. It is a struggle (*agonas*).

> The ecologists talk theory (*theoria*), we talk action (*praxi*).

Notes

1 The most labour intensive period of olive cultivation is the harvest (see Chapter Five) but Vassilikiots tend their olive groves throughout the agricultural year, trimming unnecessary branches, spraying them with pesticides and ploughing the soil with tractors at periodic intervals.

2 Palsson (1990) attempts a diachronic analysis of the symbolism of aquatic animals in Iceland. In the Icelandic past, as early as the time of settlement, and later, in the course of Icelandic history, the Icelanders' approach towards the aquatic environment was permeated by passivity, a sense of respect and lack of control. Small-scale subsistence production and patron-client labour-service contracts between landowners and landless people provided a limitation, 'a kind of ceiling', on the degree of appropriating natural resources. During that period, folk-tales, mythology and fish

symbolism, as Palsson carefully demonstrates, reflected the importance and relative power of aquatic creatures, real and imagined, in the lives of Icelandic people. But later, at the beginning of this century, a great change took place in the Icelandic attitudes towards fish and the marine environment, parallel to the advent of capitalist fishing and the commercial large-scale exploitation of aquatic resources. The older symbolic representations of fish and the sea became outdated and novel world views emerged, portraying humans as active and dominant agents and the ocean as a passive and exploitable resource.

3 Strang demonstrates that the attitude of Northern Australian cattle rangers to their environment and their animals, although 'overtly rational and economic, is also covertly affective' (1977: 166, 130). The same author also argues that Aboriginal ideas about the human environmental relationship – which represent a more symbolic and less controlling interaction with the land – have infiltrated into white Australian culture, affecting the meaning and aesthetic appreciation of white Australian mythic landscapes (Strang 2000).

4 Herzfeld (1991a) studied a conflict over archaeological conservation affecting the inhabitants of Rethemnos, a town in Crete. Rethemniots are not allowed to demolish, rebuild or modernise their old homes, just as my respondents in Vassilikos are denied control over their landed property. As Herzfeld carefully noted, in Rethemnos the homes to be conserved constitute a traditional form of dowry and the conservation restrictions touch upon several culturally significant values, being a threat to the locally portrayed need for autonomy, an intrusion of external forces into the private domestic domain, a challenge of male assertion over matrifocal property and more importantly, an obstacle to meeting the obligations of marriage. The sum of those more subtle justifications constitute a cultural exegesis more meaningful to the local protagonists than the mere calculation of the monetary value of the conserved property.

5 The environmentalists' proposition that the human species is responsible for the fate of the natural world entails elements of anthropocentrism (cf. Morris 1981: 137; O'Riordan 1976: 11, Milton 1996: 231, Norton 1991: 71–2, Kempton, Boster & Hartley 1995: 87, 95–104; see also, Hays 1987) and in this respect is not fundamentally different from the caring environmental practices of the Vassilikiots.

REFERENCES

Abram, S. 1997. Performing for tourists in rural France. In *Tourists and Tourism: Identifying with People and Places* (eds) S. Abram, J. Waldren & D.V.L. Macleod. Oxford: Berg.

Abram, S. & Waldren, J. 1997. Introduction: tourists and tourism: identifying with people and places. In *Tourists and Tourism: Identifying with People and Places* (eds) S. Abram, J. Waldren & D.V.L. Macleod. Oxford: Berg.

Abram, S., Waldren, J. & Macleod, D.V.L. (eds) 1997. *Tourists and Tourism: Identifying with People and Places*. Oxford: Berg.

Abramson, A. 2000. Mythical land, legal boundaries: wondering about landscape and other tracts. In *Land, law and environment: mythical lands, legal boundaries* (eds) Allen Abramson & Dimitrios Theodossopoulos. London: Pluto Press.

Anderson, D. & Grove, R. 1987. Introduction: the scramble of Eden: past, present and future in African conservation. In *Conservation in Africa: People, Policies and Practice* (eds) D. Anderson & R. Grove. Cambridge: Cambridge University Press.

Arapis, T. (ed.) 1992. *Thalassio Parko Zakynthou. Proypotheseis kai protaseis gia to skhediasmo, tin idhrysi kai leitourgeia tou*. Independent study, supported by WWF-Greece, Greenpeace-Greece, Sea Turtle Protection Society of Greece.

Ardener, E. 1975. The problem revisited. In *Perceiving women* (ed.) S. Ardener. London: Malaby Press.

Argyrou, V. 1996. *Tradition and modernity in the Mediterranean: the wedding as symbolic struggle*. Cambridge: Cambridge University Press.

Argyrou, V. 1997. 'Keep Cyprus clean': Littering, Pollution, and Otherness. *Cultural Anthropology* 12(2): 159–178.

Atran, S. 1990. *Cognitive Foundations of Natural History*. Cambridge: Cambridge University Press.

Atran, S. 1993. 'Whither "ethnoscience"'? In *Cognitive Aspects of Religious Symbolism* (ed.) P. Boyer. Cambridge: Cambridge University Press.

Bakalaki, A. 1997. 'Students, natives, colleagues: encounters in academia and in the field'. *Cultural Anthropology* 12(4): 502–526.

Balme, D.M. 1972. *Aristotle's De Partibus Animalium I and De Generatione Animalium I.* Oxford: Oxford University Press.

Bender, B. (ed.) 1993a. *Landscape: Politics and Perspectives* (ed.) B. Bender. Oxford: Berg.

Bender, B. 1993b. Introduction: landscape – meaning and action. In *Landscape: Politics and Perspectives* (ed.) B. Bender, 1–17. Oxford: Berg.

Berglund, E.K. 1998. *Knowing Nature, Knowing Science: An Ethnography of Environmental Activism.* Cambridge: The White Horse Press.

Berlin, B. & Berlin, E.A. 1983. Adaptation and ethnozoological classification: theoretical implications of animal resources and diet of the Aguaruna and Huambisa. In *Adaptive responses of native Amazonians* (eds) R. Hames and W. Vickers. San Francisco and New York: Academic Press.

Berlin, B. 1988. 'The chicken and the egg-head revisited: further evidence for the intellectualist bases of ethnobiological classification'. In *Ethnobiology: Implications and Applications.* Proceedings of the first International Congress of Ethnobiology.

Berlin, B. 1992. *Ethnobiological Classification.* Princeton: Princeton University Press.

Blok, A. 1981. Rams and billy-goats: a key to the Mediterranean code of honour. *Man* 16(3): 427–40.

Boissevain, J. (ed.) 1996a. *Coping with Tourists.* Oxford: Berghahn.

Boissevain, J. 1996b. Introduction. In *Coping with Tourists* (ed.) J. Boissevain. Oxford: Berghahn.

Botetzagias, I. 2001. '"Nobody does it better": intra-movement conflict concerning species conservation in Greece'. Paper presented in the ECPR 29th Joint Sessions of Workshops: Environmental politics at the local level. Grenoble, France.

Bourdieu, P. 1977. *Outline of a Theory of Practice.* Cambridge: Cambridge University Press.

Bourdieu, P. 1990. *The Logic of Practice.* Stanford: Stanford University Press.

Brandes, S. 1980. *Metaphors of Masculinity: Sex and Status in Andalusian Folklore.* Pennsylvania: University of Pennsylvania Press.

Brandes, S. 1981. Like wounded stags: male sexual ideology in an Andalusian town. In *Sexual Meanings: The Cultural Construction of Gender and Sexuality* (eds) S. Ortner & H. Whitehead, 216–239. Cambridge: Cambridge University Press.

Brandes, S. 1992. Sex roles and anthropological research in rural Andalusia. In *Europe Observed* (eds) J. Pina-Cabral & J. Campbell. London: Macmillan Press Ltd.

Brown, K. & Theodossopoulos, D. 2000. The performance of anxiety: Greek narratives of the war at Kosovo. *Anthropology Today* 16(1): 3–8.

Bulmer, R. 1967. 'Why is the cassowary not a bird? A problem of zoological taxonomy among the Karam of the New Guinea highlands'. *Man* (N.S.), 2: 5–25. Also in *Rules and Meanings* (ed.) M. Douglas. 1973. Harmondsworth: Penguin.

Bulmer, R. 1970. Which came first, the chicken or the egg-head? In *Echanges et Communication: mélanges offerts a Claude Lévi-Strauss* (eds) J. Pouillon & P. Maranda. Vol. 2. Paris: Mouton, The Haques.

Campbell, B. 2000. Animals behaving badly: indigenous perceptions of wildlife protection in Nepal. In *Natural enemies: people-wildlife conflict in anthropological perspective* (ed.) J. Knight. London: Routledge.

Campbell, J. 1964. *Honour, Family and Patronage: A Study of the Institutions and Moral Values in a Greek Mountain Community*. Oxford: Oxford University Press.

Campbell, J. 1992. Fieldwork among the Sarakatsani, 1954–55. In Europe observed (eds) J. de Pina-Cabral & J. Campbell. Oxford: Macmillan.

Cape, C. 1991. 'Turtles or tourists? Zakynthos: how tourism could assist the conservation of the Loggerhead turtle'. M.Sc. Dissertation in Conservation. University College London.

Carsten, J. 1989. 'Cooking money: gender and the symbolic transformation of means of exchange in a Malay fishing community'. In *Money and the Morality of Exchange* (eds) J. Parry & M. Bloch, 117–141. Cambridge: Cambridge University Press.

Chatty, D. & Colchester, M. (eds) 2002. *Conservation and Indigenous Mobile Peoples: Displacement, Forced Settlement and Sustainable Development*. Oxford: Berghahn.

Corbin, J.R. & Corbin, M.P. 1987. *Urbane Thought: Culture and Class in an Andalusian City*. Aldershot: Gower Publishing Company Ltd.

Cotgrove, S. 1982. *Catastrophe or Cornucopia: the environment, politics and the future*. Chichester: John Wiley & Sons.

Couroucli, M. 1985. *Les oliviers du lignage: une Grèce de tradition Vénitienne*. Paris: Maisonneuve et Larose.

Cowan, J.K. 1990. *Dance and the Body Politic in Northern Greece*. Princeton: Princeton University Press.

Cowan, J.K. 1991. 'Going out for coffee? Contesting the grounds of gendered pleasures in everyday sociability'. In *Contested Identities: Gender and Kinship in Modern Greece* (eds) P. Loizos & E. Papataxiarchis. Princeton: Princeton University Press.

Crawford, S.E. 1982. 'Person and place in Kalavasos: perspectives on social change in a Greek-Cypriot village'. Unpublished Ph.D thesis, University of Cambridge.

Croll, E. & Parkin, D. (eds) 1992a. *Bush Base: Forest Farm: Culture, Environment and Development*. London: Routledge.

Croll, E. & Parkin, D. 1992b. Cultural understanding of the environment. In *Bush Base: Forest Farm: Culture, Environment and Development* (eds) E, Croll & D, Parkin, 11–36. London: Routledge.

Danforth, L.M. 1982. *The Death Rituals of Rural Greece*. Princeton: Princeton University Press.

Danforth, L.M. 1989. *Firewalking and Religious Healing: The Anastenaria of Greece and the American Firewalking Movement*. Princeton: Princeton University Press.

Davies, D. 1994. Christianity. In *Attitudes to Nature* (ed.) J. Holm. London: Pinter Publishers.

Davis, J. 1973. *Land and Family in Pisticci*. London: Athlone Press.

Descola, P. & Palsson, G. 1996. Introduction. In *Nature and Society: Anthropological Perspectives* (eds) P. Descola & G. Palsson. London: Routledge.

Douglas, M. 1957. Animals in Lele Religious Symbolism. Africa, 27: 46–58. Also in *Implicit Meanings: Essays in Anthropology* M. Douglas. 1975. London: Routledge.

Douglas, M. 1966. *Purity and Danger*. London: Routledge.

Douglas, M. 1975. *Implicit Meanings: Essays in Anthropology*. London: Routledge.

Du Boulay, J. 1974. *Portrait of a Greek Mountain Village*. Oxford: Clarendon Press.

Du Boulay, J. 1986. Women – images of their nature and destiny in rural Greece. In *Gender and Power in Rural Greece* (ed.) J. Dubisch. Princeton: Princeton University Press.

Dubisch, J. 1986. Introduction. In *Gender and Power in Rural Greece* (ed.) J. Dubisch. Princeton: Princeton University Press.

Dubisch, J. 1991. Gender, kinship, and religion: 'reconstructing' the anthropology of Greece. In *Contested Identities: Gender and Kinship in Modern Greece* (eds) P. Loizos & E. Papataxiarchis. Princeton: Princeton University Press.

Dubisch, J. 1993. '"Foreign chickens" and other outsiders: gender and community in Greece'. In *American Ethnologist* 20(2): 272–287.

Dubisch, J. 1995a. *In a Different Place: Pilgrimage, Gender, and Politics of a Greek Island Shrine.* Princeton: Princeton University Press.

Dubisch, J. 1995b. Lovers in the field: sex, dominance, and the female anthropologists. In *Taboo: Sex, Identity, and Erotic Subjectivity in Anthropological Fieldwork* (eds) D. Kulick & M. Wilson. London: Routledge.

Durkheim, E. & Mauss, M. 1963. *Primitive Classification.* London: Cohen & West.

Ecumenical Patriarchate. 1990. *Orthodoxy and the Ecological Crisis.* A publication of the Ecumenical Patriarchate and World Wide Fund for Nature International.

Einarsson, N. 1993. All animals are equal but some are cetaceans: conservation and culture conflict. In *Environmentalism: The View from Anthropology* (ed.) K. Milton, 73–84. London: Routledge.

Ellen, R. 1986a. What Black Elk left unsaid: on the illusory images of Green primitivism. *Anthropology Today*, 2(6): 8–12.

Ellen, R. 1986b. Ethnobiology, cognition and the structure of prehension: some general theoretical notes. *Journal of Ethnobiology*, 6: 83–98.

Ellen, R. 1993. The cultural relations of classification: an analysis of Nuaulu animal categories from central Seram. Cambridge: Cambridge University Press.

Ellen, R. 1996a. Introduction. In *Redefining Nature: Ecology, Culture and Domestication* (eds) R. Ellen & F. Katsuyoshi,1–36. Oxford: Berg.

Ellen, R. 1996b. The cognitive geometry of nature: a contextual approach. In *Nature and Society: Anthropological Perspectives* (eds) P. Descola & G. Palsson. London: Routledge.

Ellen, R. & Reason, D. (eds) 1979. *Classifications in their Social Context.* London: Academic Press.

Ellen, R. & Katsuyoshi, F. (eds) 1996. *Redefining Nature: Ecology, Culture and Domestication.* Oxford: Berg.

Evans-Pritchard, E.E. 1940. *The Nuer: A Description of the Modes of Livelihood and Political Institutions of a Nilotic People.* Oxford: Oxford University Press.

Faubion, J.D. 1993. *Modern Greek Lessons: A Primer in Historical Constructivism.* Princeton: Princeton University Press.

Forde, D. (ed.) 1954. *African Worldviews: Studies in the Cosmological Ideas and Social Values of African Peoples.* Oxford: Oxford University Press.

Friedl, E. 1962. *Vassilika: A Village in Modern Greece.* New York: Holt, Rinehart and Winston.

Friedl, E. 1967. The position of women: appearance and reality. *Anthropological Quarterly* 40(3): 97–108.

Friedl, E. 1970. Field work in a Greek village. In *Women in the Field: Anthropological Experiences* (ed.) P. Golde. Berkeley: University of California Press.

Friedman, J. 1974. Marxism, Structuralism and vulgar materialism. Man (N.S.) 9: 444–469.

Galani-Moutafi, V. 1993. 'From agriculture to tourism: property, labour, gender and kinship in a Greek island village (Part One)'. *Journal of Modern Greek Studies* 11: 241–270.

Galani-Moutafi, V. 1994. 'From agriculture to tourism: property, labour, gender and kinship in a Greek island village (Part Two)'. *Journal of Modern Greek Studies* 12: 113–131.

Galaty, J.G. 1989. Cattle and cognition: aspects of Maasai practical reasoning. In *The Walking Larder* (ed.) Juliet Clutton-Brock. London: Unwin Hyman.

Gavrielides, N. 1976. 'The cultural ecology of olive growing in the Fourni Valley'. In *Regional variation in modern Greece and Cyprus: Towards a Perspective on the Ethnography of Greece* (eds) M. Dimen & E. Friedl. New York: The New York Academy of Sciences.

Geertz, C. 1973. *The Interpretation of Cultures*. New York: Basic Books.

Gefou-Madianou, D. 1992a. Introduction: alcohol commensality, identity transformations and transcendence. In *Alcohol, Gender and Culture* (ed.) D. Gefou-Madianou. London: Routledge.

Gefou-Madianou, D. 1992b. 'Exclusion and unity, retsina and sweet wine: commensality and gender in a Greek agrotown'. In *Alcohol, Gender and Culture* (ed.) D. Gefou-Madianou. London: Routledge.

Genesis. In *The New English Bible*. 1970.

Gilmore, D. 1980. *The People of the Plain: Class and Community in Lower Andalusia*. New York: Columbia University Press.

Gilmore, D. 1990. *Manhood in the Making: Cultural Concepts of Masculinity*. New Haven & London: Yale University Press.

Goddard, V. A. 1996. *Gender, Family and Work in Naples*. Oxford: Berg.

Green, S. & King, G. 1996. 'The importance of goats to a natural environment: a case study from Epirus (Greece) and South Albania'. *Terra Nova* 8: 655–58.

Green, S. & King, G. 2001. 'Seeing what you know: changing constructions and perceptions of landscape in Epirus, Northwestern Greece, 1945 and 1990'. *History and Anthropology* 12(3): 255–288.

Green, S. 1997. Interweaving landscapes: the relevance of ethnographic data on rural groups in Epirus for Palaeolithic research. In *Klithi: Palaeolithic Settlement and Quaternary Landscapes in Northwest Greece: Vol. 2, Klithi in its Local and Regional Setting* (ed.) G. Bailey. Cambridge: McDonald Institute Monographs.

Greenwood, D. J. 1976. *Unrewarding Wealth: The Commercialisation and Collapse of Agriculture in a Spanish Basque town*. Cambridge: Cambridge University Press.

Greger, S. 1988. *Village on the Plateau: Magoulas: A Mountain Village in Crete*. Warwickshire: Brewin Books.

Gudeman, S. & Rivera, A. 1990. *Conversations in Colombia: The Domestic Economy in Life and Text.* Cambridge: Cambridge University Press.

Handman, M.E. 1987. *Via kai poniria: antres kai gynaikes s' ena elliniko horio.* Athens: Ekdoseis Kastanioti.

Hannell, D. 1989. 'The Ionian islands under the British protectorate: social and economic problems'. *Journal of Modern Greek Studies* 7: 105–132.

Harries-Jones, P. 1993. Between science and shamanism: the advocacy of environmentalism in Toronto. In *Environmentalism: The View from Anthropology* (ed.) K. Milton, 43–58. London: Routledge.

Harrison, D. (ed.) 1992. *Tourism and the Less Developed Countries.* London: Belhaven Press.

Hart, L.K. 1992. *Time, Religion, and Social Experience in Rural Greece.* Lanham: Rowman & Littlefield Publishers.

Hays, S.P. 1987. *Beauty, Health and Permanence: Environmental Politics in the United States 1955–1985.* Cambridge: Cambridge University Press.

Hell, B. 1996. 'Enraged hunters: the domain of the wild in north-western Europe'. In *Nature and Society: Anthropological Perspectives* (eds) P. Descola & G. Palsson. London: Routledge.

Herzfeld, M. 1985a. *The Poetics of Manhood: Contest and Identity in a Cretan Mountain Village.* Princeton: Princeton University Press.

Herzfeld, M. 1985b. Law and custom: ethnography of and in Greek national identity. In *Journal of Modern Greek Studies* 3: 167–185.

Herzfeld, M. 1986. *Ours Once More: Folklore, Ideology, and the Making of Modern Greece.* New York: Pella Publishing Company.

Herzfeld, M. 1987. *Anthropology Through the Looking-Glass: Critical Ethnography in the Margins of Europe.* Cambridge: Cambridge University Press.

Herzfeld, M. 1991a. *A Place in History: Social and Monumental Time in a Cretan Town.* Princeton: Princeton University Press.

Herzfeld, M. 1991b. 'Silence, submission, and subversion: towards a poetics of womanhood'. In *Contested Identities: Gender and Kinship in Modern Greece* (eds) P. Loizos & E. Papataxiarchis. Princeton: Princeton University Press.

Herzfeld, M. 1992. *The Social Production of Indifference: Exploring the Symbolic Roots of Western Bureaucracy.* Chicago: The University of Chicago Press.

Hirsch, P. & O'Hanlon, M. (eds). 1995. *The Anthropology of Landscape: Perspectives on Place and Space.* Oxford: Clarendon Press.

Hirsch, P. 1995. 'Landscape: between place and space'. In *The Anthropology of Landscape: Perspectives on Place and Space* (eds) E. Hirsch & M. O'Hanlon, 1–30. Oxford: Clarendon Press.

Hirschon, R. 1978. 'Open body/closed space: the transformation of female sexuality'. In *Defining Females: the Nature of Women in Society* (ed.) S. Ardener. London: Croom Helm London.

Hirschon, R. 1983. 'Under one roof: marriage, dowry, and family relations in Piraeus'. In *Urban Life in Mediterranean Europe: Anthropological Perspectives* (eds) M. Kenny and D.I. Kertzer. Urbana: University of Illinois Press.

Hirschon, R. 1989. *Heirs of the Greek Catastrophe: The Social Life of Asia Minor Refugees in Piraeus*. Oxford: Clarendon Press.

Hunn, E. 1976. 'Towards a perceptual model of folk biological classification'. *American Ethnologist*, 3: 508–24.

Hunn, E. 1982. 'The utilitarian factor in folk biological classification'. *American Anthropologist*, 84: 830–47.

Ingold, T. 1980. *Hunters, Pastoralists and Ranchers*. Cambridge: Cambridge University Press.

Ingold, T. 1986. The Appropriation of Nature: Essays on Human Ecology and Social Relations. Manchester: Manchester University Press.

Ingold, T. (ed.) 1988. 'Introduction'. In *What is an animal?* (ed.) T. Ingold. London: Unwin Hyman. (republished by Routledge, 1994, with a new preface).

Ingold, T. 1994. 'From trust to domination: an alternative history of human-animal relations'. In *Animals and human society.* (eds) A. Manning and J. Serpell. London: Routledge.

Ingold, T. 1995. 'Building, dwelling, living'. In *Shifting Contexts* (ed.) M. Strathern, 57–80. London, New York: Routledge.

Ingold, T. 1996. Hunting and gathering as ways of perceiving the environment. In *Redefining Nature: Ecology, Culture and Domestication* (eds) R. Ellen & K. Fukui. Oxford: Berg.

Jackson, M. 1998. *Minima Ethnographica: Intersubjectivity and the Anthropological Project*. Chicago: The University of Chicago Press.

Just, R. 1994. The reformation of class. *Journal of Modern Greek Studies* 12: 37–56.

Just, R. 2000. *A Greek Island Cosmos: Kinship and Community on Meganisi*. Oxford: James Currey.

Kalligas, P.G. 1993. Oikisi stin arhaia Zakyntho. In *Oi oikismoi tis Zakynthou apo tin arhaeotita mehri to 1953*. Edited by Etaeria Zakynthiakon Spoudon. Athens: Etaireia Zakynthinon Spoudon.

Kampolis, V. 1991. *Kynigetikos Exantas*. Athens: GRAFIN-IRIS.

Karakasidou, A. N. 1997. *Fields of Wheat, Hills of Blood: Passages to Nationhood in Greek Macedonia 1870–1990*. Chicago: The University of Chicago Press.

Kemf, E. 1993. 'Tourism versus turtles'. In *Indigenous Peoples and Protected Areas: The Law of Mother Earth* (ed.) E. Kemf. London: Earthscan Publications.

Kempton, W., Boster, J. S. & Hartley, J. A. 1996. *Environmental Values in American Culture*. London: MIT Press.

Kenna, M.E. 1976a. 'Houses, fields and graves: property and ritual obligations on a Greek island'. *Ethnology* 15(1): 21–34.

Kenna, M.E. 1976b. 'The idiom of family'. In *Mediterranean Family Structures* (ed.) J. G. Peristiany. Cambridge: Cambridge University Press.

Kenna, M.E. 1990. Family, economy and community on a Greek island'. In *Family, Economy and Community* (ed.) C.C. Harris. Cardiff: University of Wales Press.

Kenna, M.E. 1993. 'Return migrants and tourism development: an example from the Cyclades. *Journal of Modern Greek Studies* 11: 60–74.

Kenna, M.E. 1995. 'Saying "no" in Greece: some preliminary thoughts on hospitality, gender and the evil eye'. In *Les amis et les autres: melanges en l'honneur de John Peristiany / Brothers and others: essays in honour of John Peristiany.* Athens: Greek National Centre of Social Research (EKKE).

Kenna, M.E. 1992a. 'Changing places and altered perspectives: research on a Greek island in the 1960s and in the 1980s'. In *Anthropology and Autobiography* (eds) J. Okely & H. Callaway. London: Routledge.

Kenna, M.E. 1992b. 'Mattresses and migrants: a patron saint's festival on a small Greek island over two decades'. In *Revitalising European Rituals* (ed.) J. Boissevain, 155–172. London: Routledge.

Kenna, M.E. 2001a. *Greek Island Life: Fieldwork in Anafi.* Amsterdam: Harwood Academic Publishers.

Kenna, M.E. 2001b. *The Social Organisation of Exile: Greek Political Detainees in the 1930s.* Amsterdam: Harwood Academic Publishers.

Knight, J. 1996. 'When timber grows wild: the desocialisation of Japanese mountain forests'. In *Nature and Society: Anthropological Perspectives* (eds) P. Descola & G. Palsson, 221–239. London: Routledge.

Knight, J. 2000a. 'Introduction'. In *Natural Enemies: People-Wildlife Conflict in Anthropological Perspective* (ed.) J. Knight. London: Routledge.

Knight, J. 2000b. 'Culling demons: the problem of bears in Japan'. In *Natural Enemies: People-Wildlife Conflict in Anthropological Perspective* (ed.) J. Knight. London: Routledge.

Konomos, K. 1979. *Zakynthos, pentakosia khronia: Ypaithros khora.* Athens: K.Michalas.

Konomos, K. 1981. *Zakynthos, pentakosia khronia: Politiki istoria* (volume A). Athens: K.Michalas.

Konomos, K. 1983. *Zakynthos, pentakosia khronia: Politiki istoria* (volume B). Athens: K.Michalas.

Konomos, K. 1985. *Zakynthos, pentakosia khronia: Politiki istoria* (volume C). Athens: K.Michalas.

Konomos, K. 1986. *Zakynthos, pentakosia khronia: Politiki istoria* (volume D). Athens: S.A.Tsupetas.

Kourtesi-Philipaki, G. 1993. I proistoriki katikisi tis Zakynthou. In *Oi oikismoi tis Zakynthou apo tin archaeotita mehri to 1953.* Edited by Etaeria Zakynthiakon Spoudon, Athens: Etaireia Zakynthinon Spoudon.

Leach, E. 1964. 'Animal categories and verbal abuse'. In *New Directions in the Study of Language* (ed.) E. H. Lenneberg. Cambridge, Mass.: MIT Press.

Leach, E. 1969. *Genesis as Myth and Other Essays.* London: Jonathan Cape Ltd.

Leontidou, L. 1995. 'Gender dimensions of tourism in Greece: employment, sub-cultures and restructuring'. In *Tourism: a Gender Analysis* (eds) V. Kinnaird & D. Hall, 74–105. Chichester: John Wiley & Sons.

Lévi-Strauss, C. 1962. *The Savage Mind.* London: Weidenfeld & Nicolson.

Lindquist, G. 2000. 'The wolf, the Saami and the urban shaman: predator symbolism in Sweden'. In *Natural Enemies: People-Wildlife Conflict in Anthropological Perspective* (ed.) J. Knight. London: Routledge.

Lison-Tolosana, C. 1966. *Belmonte de los Caballeros: A Sociological Study of a Spanish Town.* Oxford: Clarendon Press.

Loizos, P. 1975. *The Greek Gift: Politics in a Greek Cypriot Village.* Oxford: Basil Blackwell.

Loizos, P. 1981. *The Heart Grown Bitter: A Chronicle of Cypriot War Refugees.* Cambridge: Cambridge University Press.

Loizos, P. 1992. 'User-friendly ethnography?' In *Europe Observed* (eds) J. de Pina-Cabral & J. Campbell. Oxford: Macmillan.

Loizos, P. 1994. 'A broken mirror: masculine sexuality in Greek ethnography'. In *Dislocating Masculinity* (eds) A. Cornwall & N. Lindisfarne. Routledge: London.

Loizos, P. & Papataxiarchis, E. (eds) 1991a. 'Introduction: gender and kinship in marriage and alternative contexts'. In *Contested Identities: Gender and Kinship in Modern Greece* (eds) P. Loizos & E. Papataxiarchis. Princeton: Princeton University Press.

Loizos, P. & Papataxiarchis, E. (eds) 1991b. *Contested Identities: Gender and Kinship in Modern Greece.* Princeton: Princeton University Press.

Lowe, P. & Goyder. J. 1983. *Environmental Groups in Politics.* London: Allen & Unwin.

MacCannel, D. 1976. *The Tourist: A New Theory of the Leisure Class.* Berkeley: University of California Press.

MacCormack, C.P. 1980. 'Nature, culture and gender: a critique'. In *Nature, Culture and Gender* (eds) C. P. MacCormack & M. Strathern, 1–24. Cambridge: Cambridge University Press.

Macleod, D. 1999. Tourism and the globalization of a Canary island. in *Jounal of the Royal Anthropological Institute.* (N.S.) 5: 443–456.

Marcus, G.E. 1998. *Ethnography through Thick and Thin.* Princeton: Princeton University Press.

Margaritoulis, D., Dimopoulos, D. & Kornaraki, E. 1991. 'Monitoring and conservation of *Caretta caretta* on Zakynthos'. Report submitted to the EEC (Medspa-90-1/GR/28/GR/05) and WWF (project 3825).

Marvin, G. 1988. *Bullfight.* Urbana: University of Illinois Press.

Marvin, G. 2000a. 'Natural instincts and cultural passions: transformations and performances in foxhunting'. *Performance Research* 5(2): 108–115.

Marvin, G. 2000b. 'The problem of foxes: legitimate and illegitimate killing in the English countryside'. In *Natural Enemies: People-Wildlife Conflict in Anthropological Perspective* (ed.) J. Knight. London: Routledge.

McCormick, J. 1989. *The Global Environmental Movement.* London: Belhaven Press.

Milton, K. 1993a. *Environmentalism: The View from Anthropology.* London: Routledge.

Milton, K. 1993b. 'Introduction: environmentalism and anthropology'. In *Environmentalism: The View from Anthropology* (ed.) K. Milton, 1–17. London: Routledge.

Milton, K. 1996. *Environmentalism and Cultural Theory: Exploring the Role of Anthropology in Environmental Discourse.* London: Routledge.

Milton, K. 2000. 'Ducks out of water: nature conservation as boundary maintenance'. In *Natural Enemies: People-Wildlife Conflict in Anthropological Perspective* (ed.) J. Knight. London: Routledge.

Mitchell, J. 1996. 'Presenting the past: cultural tour-guides and the sustaining of European identity in Malta'. In *Sustainable Tourism in Islands and Small States: Case Studies* (eds) L. Briguglio, R. Butler, D. Harrison & W. Leal Filho. London: Pinter.

Moore, H.L. 1988. *Feminism and Anthropology*. Cambridge: Polity Press.

Moore, R.S. 1994. 'Metaphors of encroachment: hunting for wolves on a central Greek mountain'. *Anthropological Quarterly* 67: 81–88.

Morris, B. 1976. 'Whither the savage mind? Notes on the natural taxonomies of a hunting and gathering people'. *Man* (N.S.). 11: 542–57.

Morris, B. 1981. 'Changing views of nature'. *The Ecologist* 11: 130–7.

Morris, B. 1984. 'The pragmatics of folk classification'. *Journal of Ethnobiology*. 4: 45–60.

Morris, B. 1995. 'Woodland and village: reflections on the "animal estate" in rural Malawi'. *Journal of the Royal Anthropological Institute* 1: 301–15.

Morris, B. 1998. *The Power of Animals: An Ethnography*. Oxford: Berg.

Mouzelis, N. P. 1976. 'Greek and Bulgarian peasants: aspects of their sociopolitical situation during the interwar period'. *Comparative Studies in Society and History* 18(1): 85–105.

Mouzelis, N. P. 1978. *Modern Greece: facets of underdevelopment*. London: Macmillan Press.

Mylonas, D.E. 1982. *Zakynthos: istorikos, Laographikos. Kai Touristikos Odhiges*. Zakynthos: D.F. Mylonas.

Naipaul, V.S. 1987. *The Enigma of Arrival*. Harmondsworth: Penguin.

Norton, B.G. 1991. *Toward Unity Among Environmentalists*. Oxford: Oxford University Press.

O'Riordan, T. 1976. *Environmentalism*. London: Pion Limited.

Palsson, G. 1990. 'The idea of fish: land and sea in the Icelandic world-view'. In *Signifying Animals* (ed.) R. Willis, 119–133. London: Unwin Hyman.

Palsson, G. 1996. 'Constructing natures: symbolic ecology and social practice'. In *Nature and Society: Anthropological Perspectives* (eds) P. Descola & G. Palsson, 63–81. London: Routledge.

Panourgia, N. 1995. *Fragments of Death, Fables of Identity: An Athenian Anthropography*. Madison: The University of Wisconsin Press.

Papagaroufali, E. 1996. 'Xenotransplantation and transgenesis: immoral stories about human-animal relations in the West'. In *Nature and Society: Anthropological Perspectives* (eds) P. Descola & G. Palsson, 240–255. London: Routledge.

Pataxiarchis, E. 1988. 'Kinship, Friendship and Gender Relations in Two Aegean Greek Communities'. Unpublished PhD thesis, University of London.

Pataxiarchis, E. 1991. 'Friends of the Heart: Male Commensal Solidarity, Gender, and Kinship in Aegean Greece'. In *Contested Identities: Gender and Kinship in Modern Greece* (eds) P. Loizos & E. Papataxiarchis. Princeton: Princeton University Press.

Pataxiarchis, E. 1995. 'Male Mobility and Matrifocality in the Aegean Basin'. In *Brothers and Others: Essays in Honour of John Peristiany* (eds) S. Damianakos, M. Handman, J. Pitt-Rivers, G. Ravis-Giordani. Ouvrage publié avec le concours de la Maison des Sciences de l'Homme: Paris, Athens.

Papoutsopoulos, H. N. 1992. Η Εξαηερος Δηιουργα κατα τον Μεγαν Βασιλειον. Athens: Αδε?λφοτης Θεολογων ο Σωτηρ.

Parry. J. & Bloch, M. (eds) 1989. 'Introduction: money and the morality of exchange. In *Money and the Morality of Exchange* (eds) J. Parry & M. Bloch, 1–32. Cambridge: Cambridge University Press.

Pina-Cabral, J. 1986, *Sons of Adam, Daughters of Eve: The Peasant Worldview of the Alto Minho*. Oxford: Clarendon Press.

Pratt, M. 1978. *Britain's Greek Empire*. London: Rex Collings.

Price, M.F. (ed.) 1996. *People and Tourism in Fragile Environments*. Chichester: John Wiley & Sons.

Princen, T. & Finger, M. (eds) 1994. *Environmental NGOs in World Politics: Linking the Local and the Global*. London: Routledge.

Reiter, R.R. 1975. 'Introduction'. In *Toward an Anthropology of Women* (ed.) R. R. Reiter. New York: Monthly Review Press.

Richards, P. 1986. *Coping with Hunger: Hazard and Experiment in an African Rice-Farming System*. London: Allen and Unwin.

Richards, P. 1993. 'Natural symbols and natural history: chimpanzees, elephants and experiments in Mende thought'. In *Environmentalism: The View from Anthropology* (ed.) K. Milton, 144–159. London: Routledge.

Richards, P. 1996. 'Agrarian creolization: the ethnobiology, history, culture and politics of West African rice'. In *Redefining Nature: Ecology, Culture and Domestication* (eds) R. Ellen & F. Katsuyoshi, 291–318. Oxford: Berg.

Richards, P. 2000. 'Chimpanzees as political animals in Sierra Leone'. In *Natural Enemies: People-Wildlife Conflict in Anthropological Perspective* (ed.) J. Knight. London: Routledge.

Ritvo, H. 1987. *The Animal Estate*. Cambridge, MA: Harvard University Press (Harmondsworth: Penguin 1990).

Ritvo, H. 1994. Animals in nineteenth-century Britain: Complicated attitudes and competing categories. In *Animals and Human Society* (eds) A. Manning and J. Serpell. London: Routledge.

Roma, D.A. 1967. *Periplous* (1570–1870): *O Soprakomitos*. 2 Volumes. Athens: Estia.

Roma, D.A. 1971. *Periplous* (1570–1870): *To rempelio ton popolaron*. 2 Volumes. Athens: Estia.

Roma, D.A. 1973. *Periplous* (1570–1870): *O thrinos tis kantias*. 2 Volumes. Athens: Estia.

Roma, D.A. 1975. *Periplous* (1570–1870): *O Kontes*. 2 Volumes. Athens: Estia.

Roma, D.A. 1980. *Periplous* (1570–1870): *Antatzio kai fouga*. 2 Volumes. Athens: Estia.

Sahlins, M. 1976. Culture and practical reason. Chicago: University of Chicago Press.

Sakkos, S.K. 1973. *Vasileiou kaisareias tou Megalou Apanta Erga: Exaimeros*. Thessalonika: Paterikai ekdhosers Grigorios o Palamas.

Salamone, S.D. & Stanton, J.B. 1986. 'Introducing the nikokyra: identity and reality in social process'. In *Gender and Power in Rural Greece* (ed.) J. Dubisch. Princeton: Princeton University Press.

Salvator, L. 1904. *Zante: Allgemeiner Theil*. Prague: Druck and Verlag von Heiner. Mercy Sohn.

Sant Cassia, P. & Bada, C. 1992. *The Making of the Modern Greek Family: Marriage and Exchange in 19th Century Athens*. Cambridge: Cambridge University Press.

Sant Cassia, P. 1982. 'Property in Greek Cypriot Marriage Strategies 1920–1980'. *Man* 17: 643–63.

Selwyn, T. (ed.) 1996a. *The Tourist Image: Myths and Myth Making in Tourism*. Chichester: John Wiley & Sons.

Selwyn, T. 1996b. 'Introduction'. In *The Tourist Image: Myths and Myth Making in Tourism* (ed.) T. Selwyn. Chichester: John Wiley & Sons.

Seremetakis, N.C. 1991. *The Last Word: Women, Death, and Divination in Inner Mani*. Chicago: The University of Chicago Press.

Serpell, J. 1986. *In the Company of Animals*. Oxford: Blackwell.

Serpell, J. 1989. 'Pet-keeping and animal domestication: a reappraisal'. In *The Walking Larder* (ed.) Juliet Clutton-Brock. London: Unwin Hyman.

Serpell, J. and Paul, E. 1994. 'Pets and the development of positive attitudes to animals'. In *Animals and Human Society* (eds) A. Manning and J. Serpell. London: Routledge.

Schama, S. 1995. *Landscape and Memory*. London: Harper Collins.

Sidirokastriti, M. 1993. *Zakynthos: History, Art, Folklore, Tour*. Athens: Editions Palmette.

Slaughter, C. & Kasimis, C. 1986. 'Some social-anthropological aspects of Boeotian rural society: a field report'. *Byzantine and Modern Greek Studies* 10: 103–160.

Sordinas, A. 1993. *Lithina ergleia proimotatis typologias sti Zakyntho. In Oi oikismoi tis Zakynthou apo tin arkhaiotita mekhri to 1953*. Athens: Etaireia Zakynthinon Spoudon.

St Basil (the Great). *Homilies on the Hexaemeron*. Modern Greek translation by Sakkos, S.K. 1973. *Vasileiou Kaisareias tou Megalou Apanta Erga: I Exaimeros*. Thessalonika: Paterikai ekdhoseis 'Grigorios o Palamas'. English translation by Sister Agnes Clare Way, C.D.P. (translator) 1963. Exegetic Homilies. Washington: The Catholic University of America Press.

Stewart, C. 1991. *Demons and the Devil*. Princeton: Princeton University Press.

Stewart, C. 1994. 'Syncretism as a dimension of nationalist discourse in modern Greece'. In *Syncretism and Anti-syncretism* (eds) C. Stewart & R. Shaw. London: Routledge.

Stewart, C. & Shaw, R. 1994. 'Introduction'. In *Syncretism and Anti-syncretism* (eds) C. Stewart & R. Shaw. London: Routledge.

Stott, M.A. 1985. 'Property, labor and household economy: the transition to tourism in Mykonos, Greece'. *Journal of Modern Greek Studies* 3(2): 187–206.

Strang, V. 1997. *Uncommon Ground: Cultural Landscapes and Environmental Values*. Oxford: Berg.

Strang, V. 2000. 'Not so black and white: the effects of aboriginal law on Australian legislation'. In *Land, Law and Environment: Mythical Lands, Legal Boundaries* (eds) Allen Abramson & Dimitrios Theodossopoulos, 59–77. London: Pluto Press.

Strathern, M. 1980. 'No nature, no culture: the Hagen case'. In *Nature, Culture and Gender* (eds) C. P. MacCormack & M. Strathern, 174–222. Cambridge: Cambridge University Press.

Strathern, M. 1987. 'Introduction'. In *Dealing with Inequality: Analysing Gender Relations in Melanesia and Beyond* (ed.) M. Strathern. Cambridge: Cambridge University Press.

Strathern, M. 1988. *The Gender of the Gift: Problems with Women and Problems with Society in Melanesia*. Berkeley: University of California Press.

Sutton, D. 1994. '"Tradition and modernity": Kalymnian construction of identity and otherness'. In *Journal of Modern Greek Studies* 12: 239–60.

Sutton, D. 1996. 'Explosive debates: dynamite, tradition, and the state'. *Anthropological Quarterly* 69(2), 66–78.

Sutton, D. 1998. *Memories Cast in Stone: The Relevance of the Past in Everyday Life*. Oxford: Berg.

Tambiah, S.J. 1969. 'Animals are good to think and good to prohibit'. *Ethnology* 7: 423–59.

Tapper, R.L. 1988. 'Animality, humanity, morality, society'. In *What is an Animal?* (ed.) T. Ingold. London: Unwin Hyman.

Theodossopoulos, D. 1997a. 'Turtles, farmers and 'ecologists': the cultural reason behind a community's resistance to environmental conservation'. *Journal of Mediterranean Studies* 7(2): 250–67.

Theodossopoulos, D. 1997b. '"What use is the turtle?": cultural perceptions of land, work, animals and "ecologists" in a Greek farming community'. Unpublished Ph.D. Thesis, University of London.

Theodossopoulos, D. 1999. 'The Pace of the Work and the Logic of the Harvest: Women, Labour and the Olive Harvest in a Greek Island Community'. *Journal of the Royal Anthropological Institute* (N.S.) 5(4): 611–626.

Theodossopoulos, D. 2000. 'The land people work and the land the 'ecologists' want: indigenous land valorisation in a rural Greek community threatened by conservation law'. In *Land, Law and Environment: Mythical Lands, Legal Boundaries* (eds) Allen Abramson & Dimitrios Theodossopoulos, 59–77. London: Pluto Press.

Theodossopoulos, D. 2002. 'Environmental conservation and indigenous culture in a Greek island community: the dispute over the sea turtles'. In *Conservation and Indigenous Mobile Peoples: Displacement, Forced Settlement and Sustainable Development* (eds) D. Chatty & M. Colchester. Oxford: Berghahn.

Thomas, K. 1983. *Man and the Natural World: Changing Attitudes in England 1500–1800*. Harmondsworth: Penguin.

Thucydides. *History of the Peloponnesian War: Books I & II*. With an English translation by C. Forster Smith (1919). London: William Heinemann Ltd.

Tilley, C. 1994. *A Phenomenology of Landscape*. Oxford: Berg.

Toren, C. 1989. 'Drinking cash: the purification of money through ceremonial exchange in Fiji'. In *Money and the Morality of Exchange* (eds) J. Parry & M. Bloch, 142–64. Cambridge: Cambridge University Press.

Toubis, M. 1991. *Zakynthos Today and Yesterday*. Athens: Toubis Graphic Arts.

Tzannatos, Z. 1989. 'Women's wages and equal pay in Greece'. *Journal of Modern Greek Studies* 7: 155–169.

Urry, J. 1990. *The Tourist Gaze: Leisure and Travel in Contemporary Society*. London: Sage.

Van der Ploeg, J.D. & A. Long. 1994. *Born from Within: Practice and Perspective of Endogenous Rural Development*. Van Gorum: Cip-Data Koninklijke Bibliotheek.

Waldren, J. 1996. *Insiders and Outsiders: Paradise and Reality in Mallorca*. Oxford: Berghahn.

Waldren, J. 1997. 'We are not tourists – We live here'. In *Tourists and Tourism: Identifying with People and Places* (eds) S. Abram, J. Waldren & D.V.L. Macleod . Oxford: Berg.

Way, A.C. (translator) 1963. *Exegetic Homilies*. Washington: The Catholic University of America Press.

Welz, G. 1999. 'Reflexive traditionalists: return migrants as small entrepreneurs in the economic culture of tourism'. Paper presented at the conference 'Cypriot Society into the New Millennium: Globalisation and Social Change'.

White, L. 1968. *Machina ex Deo: Essays on the Dynamism of Western Culture*. Massachusetts: MIT Press.

Willis, R. 1974. *Man and Beast*. London: Hart-Davis, MacGibbon.

Willis, R. 1990. 'Introduction'. In *Signifying Animals* (ed.) R. Willis, 1–24. London: Unwin Hyman.

Worster, D. 1977. *Nature's Economy: A History of Ecological Ideas*. Cambridge: Cambridge University Press.

Xenopoulos, G. 1936. *Megali Agapi*. Athens: Adherfoi Vlassi (1984).

Xenopoulos, G. 1945. *O Popolaros*. Athens: Oi filoi fou vivliou.

Xenopoulos, G. 1959a. *Stella Violanti*. Athens: Ekdhoseis Biris.

Xenopoulos, G. 1959b. *O Kokkinos Vrakhos*. Athens: Ekdhoseis Biris.

Xenopoulos, G. 1984. *Tereza Varma-Dhakosta*. Athens: Adherfoi Vlassi.

Yanagisako, S. & Delaney, C. 1995. 'Naturalizing power'. In *Naturalizing Power: Essays in Feminist Cultural Analysis* (eds) S. Yanagisako & C. Delaney. New York: Routledge.

Zarkia, C. 1996. 'Philoxenia receiving tourists – but not guests – on a Greek island'. In *Coping with Tourists* (ed.) J. Boissevain. Oxford: Berghahn.

Zois. L. 1963. *Lexikon Istorikon kai Laographikon Zakynthou*. Volumes A&B. Athens: Ethnikon Typographeion.

Titles of magazines about hunting, dogs and guns:

Κυνηγεσία και Κυνοφιλία.

Κυνηγετικά Νέα : Μηνιαίο κυνηγετικό περιοδικό του χθες και του σήμερα.

Κυνήγι & Σκοποβολή.

Κυνήγι και όχι μόνο. Περιοδικό κυνηγετικού – σκοπευτικού – φυσιολατρικού & κυνοφιλικού περιεχομένου.

Κυνηγός & Φύση. Το σύγχρονο περιοδικό για τους λάτρες της δράσης.

INDEX

Abram, S., 50, 63
Abram and Waldren, 50, 167
Abram, Waldren and Macleod, 50
Abramson, A., 167
agonas; see struggle
agriculture, 2, 10–11, 20–1, 29, 31–3, 38,
    45–6, 49–65, 67–87, 165–8;
    neglect of, 20, 50–6;
    compatibility with tourism, 10, 21, 49–65,
    94, 162–3;
    and security, 53, 56
Albanians, 77, 85
Anderson and Grove, 2
animal classification, 11, 112, 121–35, 138n,
    170–1
animal husbandry, 10–11, 89–109, 168–9
animal usefulness, 10, 22, 89–90, 92, 95,
    111–38, 153
animals, 10–11, 89–109, 111–38, 168–74,
    see also, hunting;
    domestic, 10–11, 89–109, 168–70;
    wild, 11, 111–38, 169–74;
    flocks of, 96–100;
    attitudes to, 168–74;
    punishing, 10–11, 90, 103–5, 108, 117,
    168–9;
    as members of households, 10, 22, 89–90,
    105–9, 117, 153, 159, 168–70;
    outside the context of care and order, 159,
    169
anthropocentrism, 11–12, 127, 129–30,
    134–5, 138n, 168–74

Ardener, E., 175
Argyrou, V., 3, 4, 6, 8, 30, 49, 59, 77, 162
aristocrats, 18–19, 26n, 31, 33, 142;
    see also nobiloi
Aristotle, 121, 130, 134, 137n, 138n
Atran, S., 134, 170

Bakalaki, A., 6, 7
Bender, B., 166–7
Berglund, E.K., 3, 4
Berlin, B., 135, 138n
birds of prey, 114–5, 125–6, 136n
Blok, A., 91
Boissevain, J., 22, 42, 49–50, 54, 63, 65n
Botetzagias, I., 7, 8, 23, 27n
Bourdieu, P., 3, 67, 84, 162, 164, 174
Brandes, S., 67–8, 74, 76, 78, 91
Bulmer, R., 129–30, 134, 138n, 170

Campbell, B., 23
Campbell, J., 3, 6, 25, 68, 91, 107, 175
care and order, 11, 100–5, 106–8, 133, 153,
    155, 159, 169–70
Carsten, J., 164
cattle, 61, 90, 92–3, 98, 172
Chatty and Colchester, 2
chickens, 52, 90, 93–4, 104–5, 114, 116–7,
    119, 152–5
conservation; see environmental conservation
Corbin and Corbin, 68
Cotgrove, S., 4, 12n
Couroucli, M., 69, 74, 86n

Cowan, J.K., 3, 49, 64n, 67, 74, 82
Croll and Parkin, 3, 171

Danforth, L.M., 3, 59, 115
Davies, D., 170
Davis, J., 29–30, 42, 46, 165
Descola and Palsson, 3, 175
dogs, 90, 92, 107, 109n, 118, 126–7, 146, 151, 154,
    see also hunting dogs
dolphins, 22, 120, 123
donkeys, 61, 95–6, 116, 126
Douglas, M., 128–30, 138n, 171–2
dowry,39–40, 82, 181n
Du Boulay, J., 3, 10, 20, 22, 29–30, 33, 35, 39–40, 46, 49, 51, 58–9, 62, 68, 77, 83, 86, 89–90, 104–8, 153, 162–3, 165, 168
Dubisch, J., xii, 3, 6, 49, 59, 68, 83, 86, 108, 162
ducks, 94
Durkheim and Mauss, 134

ecologists; *see* environmentalists
Einarsson, N., 2, 159
Ellen, R., 3, 134, 138n, 159, 170–2, 175
environment, 1–3, 9–12, 12n, 13–16, 39, 43, 49, 51, 59, 60, 64, 67, 90, 100–1, 105, 108, 118, 121, 132–4, 140, 156, 161–177
environmental conservation, 1–3, 4–5, 7–8, 12n, 22–6, 43–6, 106, 135, 141, 151–2, 159, 161, 164, 169, 174–6
environmental politics, 4–5, 7–8, 22–6, 30, 43–5, 46–7, 151–2, 159, 164, 174–6
environmentalism, 4–5, 7–8, 11–12, 12n, 22–6, 106, 135, 138n, 139–40, 151–2, 159, 174–6
environmentalists, 1–3, 4–8, 11–12, 12n, 14, 21, 23–5, 106, 109, 113–4, 133, 135, 136n, 139, 151–2, 159, 161, 164, 166, 174–6
environmental NGOs, 4, 12n, 23–24
Evans-Pritchard, E.E., 99, 171–2

farming; *see* agriculture
Faubion, J.D., 3, 77, 157
feudalism, 18–20, 31–5, 69–74, 165, 142–3
fishing, 22, 120, 177n
foxes, 116, 126, 136n
Friedl, E., 3, 6, 49, 59, 68, 74, 83, 84, 93, 107–8, 162, 168

Galani-Moutafi, V., 41, 50, 68, 74, 85, 86n, 108, 168
Geertz, C., 142, 175
geese, 94, 126
Gefou-Madianou, D., 29–31, 39–40, 52, 68, 74, 157
gendered division of labour, 10, 67–87, 163–4
Gilmore, D., 67
goats, 90–2, 95–6, 98–9, 103–4
Goddard, V.A., 77, 85
Green, S., 3, 167
Green and King, 91, 167
Greenwood, D.J., 54, 63
Greger, S., 53, 74, 78, 91, 167
Gudeman and Rivera, 62, 83, 163

Handman, M.E., 6, 33, 35, 47n, 75, 86n, 105, 108, 156, 158, 168
Hannell, D., 19, 26n, 27n, 31, 69
Harries-Jones, P., 4, 12n
Harrison, D., 49
Hart, L.K., 3, 35, 40, 49, 59, 74, 86n, 107, 108, 162, 167, 168
Hays, S.P., 12n, 177n
Hell, B., 156
Herzfeld, M., 3, 5, 6, 19, 22, 24, 25, 29, 30, 32, 33, 42, 43, 80, 140, 157, 159, 165, 169, 173, 174, 175, 177n
Hexaemeron, 112, 121–35, 136n, 137n, 138n
Hirsch, P., 167
Hirsch and O'Hanlon, 166
Hirschon, R., 3, 46, 68, 74, 83, 85, 86, 108, 163
horses, 15, 61, 95, 104, 126, 173
Hunn, E., 138n
hunters, 2, 11, 14, 139–60, 169, 173
hunting, 2, 11, 14, 16, 119, 132, 135, 139–160, 175
hunting guns, 11, 141–5, 148, 151–2, 156
hunting dogs, 92, 109n, 146, 151, 154

Ingold, T., 3, 167, 169–71
inheritance, 30–2, 39–41

Jackson, M., 84
Just, R, 3, 6, 30, 40, 49, 51, 62, 74, 77, 86n, 87n, 96, 163

Karakasidou, A. N., 3
Kemf, E., 12n
Kempton, Boster and Hartley, 177
Kenna, M.E., 3, 6, 7, 41, 49, 51, 59, 62, 76,

85, 86n, 87n, 105, 143, 162, 163, 167
Knight, J., 2, 22, 60, 118
Konomos, K, 14, 17–19, 26n, 27n

Lamarck, 127
land, 1–3, 9–11, 14, 18–21, 22, 24, 29–47,
    49–52, 54–8, 62–4, 69–70, 96–8, 100,
    142–3, 148–9, 153–4, 162–3, 165–8;
    acquisition, 33–8;
    value of, 9, 29–30, 38–42, 44–5, 46–7,
    165–7;
    deserted, 17, 33–4;
    clearing, 33–4, 47n, 49, 61
landless tenants, 9, 18–20, 31–5, 37–9, 46,
    68–73, 83, 97, 143–4, 165–6;
    see also semproi
landscape, 13–16, 69, 166–7, 177n;
    empty, 13–16, 166–7
law, 23–24, 32–33, 43, 87n, 146, 149, 164,
    175
legislation, 9, 25, 33, 44, 47n, 144, 156, 172
Leach, E., 128–30, 138n
Leontidou, L., 50, 64n
Lévi-Strauss, C., 129, 134
Lindquist, G., 22, 116
Lison-Tolosana, C., 29–30, 64n, 165
Loizos, P., 3, 6, 25, 29, 30, 41, 46, 62, 83, 85,
    86, 108, 157, 163, 165, 175
Loizos and Papataxiarchis, 41, 68, 83, 86, 108,
    164
Lowe and Goyder, 4

MacCannel, D., 7, 63, 65n
MacCormack, C.P., 3
Macleod, D., 50
Marcus, G.E, 6
Margaritoulis, D., 24
Marvin, G., 22, 92, 93, 99, 111, 116, 118,
    133, 139, 158, 169
middle-class, 4, 6–7, 13–16, 18–19, 26n,
    31–4, *see also* popolaroi
Milton, K., 3, 4, 7, 10n, 176, 177n
Mitchell, J., 54, 63
Moore, H.L., 80
Moore, R.S., 116, 132, 156–7, 159n, 169
Morris, B., 138n, 170, 171–3, 177n
Mouzelis, N. P., 25, 27n

Naipaul, V.S., 14
nobiloi, 18–19, 26n, 31
Norton, B.G., 4, 12n, 177n

O'Riordan, T., 12n, 177n
olive harvest, 7, 10, 67–8, 74–7, 78–80, 83–5,
    163
olive trees, 17, 36, 56, 61, 64, 69–70, 73,
    74–6, 100, 109n, 150
order; *see* care and order

Palsson, G., 168, 171, 176n, 177n
Panourgia, N., 3, 6
Papagaroufali, E., 170
Papataxiarchis, E., 11, 156, 157, 160n
Parry and Bloch, 164, 174
Patrilocality, 41, 82
pigs, 94
Pina-Cabral, J., 12n, 29–30, 46, 62, 64n, 75,
    81, 87n, 163, 165, 168
Plato, 136n, 172–3
popolaroi, 18–19, 26n, 31
predators, 93, 104, 111, 116, 117, 120, 127,
    168, 169
Price, M.F., 49–50
Princen and Finger, 7, 12n

rabbits, 52–3, 94, 104–5, 116–19, 153–5
reciprocity, 22, 57, 64n, 100, 105, 168, 171
Reiter, R.R., 80
Richards, P., 22, 25, 171
Ritvo, H., 170, 173
robin, 113, 135n, 136n, 143, 160n
Roma, D.A, 18–19, 26n, 27n, 40, 64n, 142

Sahlins, M., 161–2, 174
Salamone and Stanton, 68, 83, 86, 108
Salvator, L, 69, 87n
Sant Cassia, P., 30, 33
Sant Cassia and Bada, 3
Schama, S., 166
seal, 22, 114, 120, 123, 132, 135, 136n, 172n
self-interest, 46, 83, 85, 108, 163, 164
self-sufficiency, 10, 29, 46, 50–3, 57, 62, 64n,
    73–4, 77, 83, 85, 87n, 91–2, 94, 100,
    104–5, 155, 163, 164, 165
Selwyn, T., 49, 63, 65
sembremata, sempries, 20, 31–6, 38, 47,
    69–5, 87n, 97, 100, 148, 165, 109n
sembroi, 9, 15, 18–20, 26n, 30–4, 36, 46, 70,
    73, 143, 165
Seremetakis, N.C., 3, 30, 39, 75, 78, 160n
Serpell, J., 170, 172–3
Serpell and Paul, 172
sheep, 15, 56, 73, 90–2, 96–100, 103, 107,
    116, 127, 149

Sidirokastriti, M., 17, 26n
St Basil (the Great), 11, 112, 121–35, 136n,
    137n, 138n, 170
Stewart, C., 3, 30, 62, 77, 134, 163
Stewart and Shaw, 138n
Stott, M.A., 41, 55
Strang, V., 165, 166–7, 171, 177n
Strathern, M., 3, 10, 68, 84–6, 163
struggle, 10, 19, 21, 32, 38, 44, 47, 49–51,
    53, 55, 57–64, 80, 86, 102, 106, 109,
    117–18, 156, 159, 161–9, 171, 174–6
Sutton, D., 3, 54, 144, 147, 158, 159, 160n

Tambiah, S.J., 129
Tapper, R.L., 170
taxonomy; *see* animal classification
tenants; *see* landless tenants
Thomas, K., 170, 173
Thucydides, 17
Tilley, C., 166–167
Toren, C., 164
tourism, 1–2, 9–11, 20–3, 29–30, 38, 41–6,
    49–65, 73, 85, 94–5, 161–9, 174
turkeys, 93–6

turtledoves, 14, 119–20, 135, 142–4, 146–54
turtles, v, 2, 12n, 22–4, 113–14
Tzannatos, Z., 87n

Urry, J., 63, 65n

Van der Ploeg, J.D. and Long, 93, 98
Venetians, 17–21, 26n, 31, 34, 40, 47n, 69,
    83, 86n, 156

Waldren, J., 14, 55, 63, 69, 74, 75, 78
Welz, G., 50
White, L., 170
Willis, R., 4, 170–2
wolves, 116, 126, 136n
Worster, D., 12n, 170

Xenopoulos, G., 26n, 27n

Yanagisako and Delaney, 152, 159

Zarkia, C., 22, 42, 50
Zois. L., 18